MIND PENETRATION

BOOKS PREVIOUSLY PUBLISHED

by Dr. Haha Lung

The Ancient Art of Strangulation (1995)
Assassin! Secrets of the Cult of Assassins (1997)
The Ninja Craft (1997)
Knights of Darkness (1998)
Lost Fighting Arts of Vietnam (2006)
Mind Control (2006)

Written with Christopher B. Prowant

Shadowhand: Secrets of Ninja Taisavaki (2000)
The Black Science: Ancient and Modern Techniques of Ninja Mind Manipulation (2001)

Written as "Ralf Dean Omar"

Death on Your Doorstep: 101 Weapons in the Home (1993)
Prison Killing Techniques: Blade, Bludgeon & Bomb (2001)

Written as "Dirk Skinner"

Street Ninja: Ancient Secrets for Mastering Today's Mean Streets (1995)
X-Treme Boxing: Secrets of the Savage Street Boxer (2002)
    with Christopher B. Prowant

MIND PENETRATION

The Ancient Art of Mental Mastery

DR. HAHA LUNG

CITADEL PRESS
Kensington Publishing Corp.
www.kensingtonbooks.com

To Naomi Belle Shifferly:

Mother to Crystal & Jet;

Daughter to James & Agnes;

Sister to James, Nancy Ann, & Rick;

Grandmother to Kaleel, Kazaria & Pooter.

Caregiver.

CITADEL PRESS BOOKS are published by

Kensington Publishing Corp.
850 Third Avenue
New York, NY 10022

First printing: November 2007

10 9 8 7 6 5 4 3 2 1

Printed in the United States of America

Library of Congress Control Number: 2007929067

ISBN-13: 978-0-8065-2852-6
ISBN-10: 0-8065-2852-4

CONTENTS

INTRODUCTION
"How to Be Outstanding in Your Field"

"Prudent men are wont to say—and this not rashly or without good ground—that he who would foresee what has to be, should reflect on what has been, for everything that happens in the world at any time has a genuine resemblance to what happened in ancient times. This is due to the fact that the agents who bring such things about are men, and men have, and always have had, the same passions, whence it necessarily comes about that the same effects are produced."
—Machiavelli, *Discourses*

Mind control and manipulation, brainwashing—enquiring minds just can't seem to get enough. Nor should they.

The Black Science (as such mental machinations and underhanded maneuverings are collectively known) is an ever-expanding, always evolving field . . . a battlefield! A battlefield where we are all combatants, whether we like it or not.

Every day we sally forth unsuspecting, like babes into the woods, sheep to the slaughter, out into a world of potential "land mines" and "booby-traps," where ruthless enemies to our will and wallet lurk behind every seemingly innocent playbill and billboard, all designed to seize a hold on our attention—both consciously and, better yet, subconsciously. And once they've fixed their grip, they then insidiously insinuate their twisting tentacles deeper, ever deeper into our minds—catering to our existing desires and fears and, where necessary or just when possible—fostering such fears and desires within us—uncomfortable and unruly emotional weaknesses they will then be more than happy to satiate . . . but always at a price.

This is the battlefield of everyday life, where words are weapons, unchecked emotions are weaknesses for a wily foe to exploit, and where the shields of rationalization, repression, and regret we hastily throw up to protect ourselves from mental domination and ego devastation are almost always too little, too late.

Then there's that other kind of field . . . the farmer's field, where the rule of nature—especially human nature—is: You reap what you sow. Call it karma, kismet, fate, or just the "big payback"; sooner or later the chickens come home to roost. Shit happens.

But while the farmer all too often finds his future in the merciless talons of raven Fate, that is not to say he is a victim—never that!

No, the wise farmer trusts if he puts a kernel of corn into the soil, beans will not sprout instead. Likewise he knows that any seeds planted on purpose today (always in ground prepared beforehand, by the way)—in the hopes of reaping a better tomorrow—must be diligently nurtured, with an ever-vigilant eye to safeguarding our all-too-susceptible seedlings against sundry pests, vermin, and perennial thieves, until the time is ripe for harvesting—for reaping what we have sown.

Battlefield master Miyamoto Musashi encouraged us to "learn the ways of all professions."

A renowned warrior admonishing us to study the way of the farmer? That's because Musashi realized that in so many ways the killing field of the warrior is not that far removed from the tilling field of the farmer.

Like the farmer, the warrior must pick his field carefully. He must know the lay of the land—how the cooling water flows down the valley, how the hot wind whips across the flatlands. He must prepare beforehand for his time spent living—and dying—in his field. And he must study the signs—all around him—to determine the best time to beat his plowshare into a sword, even as his farming cousin knows when to beat his sword into a plowshare.

As with the farmer, so with the warrior. There is little room for error. Nature seldom gives out second-place prizes. If the warrior tends to his field diligently, then, like the farmer, he will truly reap what he has sown.

In this way one achieves respect, perhaps even glory, and is acclaimed by his fellows to be "outstanding in his field."

But what if you are merely . . . out there, standing, in your field? Then just as surely the winds and the floods and the crows—both human and sky-born—will come to rape and pillage your field.

And in the end, because you failed to prepare your field beforehand, failed to study the signs, failed to strike when the time for harvesting victory was upon you, in the end, you win a third kind of field. . . .

Potter's Field . . . where they bury the lax, the lost, and the losers.

Black Science: Generic use, any strategy, tactic, or technique used to undermine a person's ability to reason and respond for themselves.

The term was first coined in recognition of the contribution of Dr. C. B. Black to the field of Aberrant Anthropology.

 *Part I*

WHO CAN YOU TRUST?

"We are all disposed, more or less, to be tyrannized by preconceptions and first impressions, and to refuse to see more than may be seen at a glance, or, perhaps, to see more than we want to see."
—Ikbal Ali Shah, 1938

The human body and brain have weaknesses. The bad news is everyone of us—including you and me—are susceptible to these human failings, foibles, and faux pas.

The good news is that our enemies are just as susceptible to these gaffes, greeds, and gaping holes in their mental defenses. This at least gives us a fighting chance.

Down through the centuries various Black Science practitioners and cadre—from individual charlatans, schemers, political and fanatical snipes, and skulduggerists, to secret societies with all color of hidden agenda—every possible failing of humanity has been explored—and then exploited!—to the meteoric rise of some, the pitiful downfall of others. We will study and master these collections and catalogs of human failings in coming chapters.

So if human beings are so fallible, who can we trust? Well, we already know we can't trust our enemies. After all, that's why the untrustworthy and underhanded bastards are our enemies in the first place.

I know what you're thinking: Since these weaknesses are inherent in all humans, doesn't that mean, if we pay close enough attention, we will be able to read our enemies like a comic book?

1

You catch on fast.

But on the same note, does that mean I can't even trust my closest friends and family, since my enemies might use them and their potential weaknesses to get to me?

Ah, sad but true.

Can I even trust myself from being distracted and led astray by wily enemies who maybe study the Black Science more than I do?

Quite possible. But not predetermined. Your future peace of mind, as well as your own personal and financial safety—as well as that of your loved ones—is still in your hands and not yet under the tyrannical thumb of your enemies, provided you master your own potentially distorted perceptions and skewed perspectives before your enemies learn to turn these weaknesses against you.

Having mastered this initial step in the Black Science, you will then go on to learn both righteous appreciation for, and ruthless application of, time-honored Mind-Fist techniques guaranteed to guard your own mental keep while arming you to lay siege to your enemies' mind-castle!

1

Perception and Perspective

*"For the professional deceiver [magicians, spies, confidence men]
the problem is not one of moral rectitude but rather of social
comprehension. He or she must understand the seen-but-
unnoticed features of everyday life."*
—Lyman & Scott, 1989:176

Who hasn't heard the story of "The Three Blind Men and the Elephant"?

Three blind men chance upon an elephant. The first blind man runs his hand down the length of the elephant's trunk and declares, "An elephant is like a tree!"

The second runs his hands along the side of the elephant, disagreeing, "No, an elephant is like a wall."

The third blind man, feeling the beast's tail disputed his two fellows, "You are both wrong, an elephant is like a serpent!"

Could it be they were all three right? Yes. First, because their perceptions gave them correct readings. And, second, they drew logical conclusions based on their individual perspectives, i.e., where they happened to be standing.

Of course we know all three also saw incompletely because of their limited perceptions and perspective.

This is the same thing that can and does happen to each and every one of us every day: Our perceptions and our perspectives lie to us or, at the very least, give us limited information to go by, prompting us to "fill in the blanks."

Filling in the blanks is human nature. We strive for completeness. The fancy word for this is "gestalt."*

That's why two people can witness the same car accident at the same intersection, albeit from different corners, and subsequently come up with two different versions of the accident.

In the same way, where you happen to be "standing"—literally and figuratively—at any given time in your life, coupled with your personal perceptions that are filtered through all manner of drama and trauma you're carrying with you from childhood, not to mention your adult prides and prejudices, and/or whether or not you "got some" last night—can all distort how we perceive a particular event, word, or action—our own as well as the words and actions of others.

Savvy con men, cult leaders, and politicians factor these two variables—perception and perspective—into account before spinning their spiels in our direction. Having mastered the Black Science, these professional mind magicians know how to make our eyes follow their right hand while their left is picking our pocket. They know how to say just the right word to catch and keep our attention. They know how to stand—in order to project just the right kind of body language to make themselves and their offer more attractive to us.

Most important, they know the deck is already stacked in their favor since most of us remain blissfully ignorant of the fact we can't trust our own perceptions and perspectives to safely carry us through the day unmolested.

I know what you're thinking: This kind of mind manipulation crap only applies to other people, to weak-minded people. You think you are too smart, too sharp to fall for con games, cult lures, hypnotism, or "weapons of mass destruction."

"All my senses are in perfect working order," you maintain. You see what you see. You hear what you hear. And you can trust your memory 100 percent. . . . Let's test that claim, shall we?

A BODY TEST

Ask someone to describe a spiral staircase. It will be impossible for them to do this without twirling their finger in the air. Even when you telephone

---

*Gestalt, as it applies to the Black Science, is discussed more fully in my books *Mind Manipulation* (Citadel, 2002) and *Mind Control* (Citadel, 2006).

someone and ask him or her to describe a spiral staircase, all alone at the other end of the line, your victim will still be twirling his or her finger in the air!

A simple game, you say? Yes, but one with a very important lesson: Our body has a mind of its own.

From something as simple as being unable to describe a spiral staircase without twirling your finger in the air, to those beads of sweat on your brow that just lost you that $10,000 pot by ruining your bluffing at Texas Hold'em, to the blood flow quickly heading south in response to that big bootied blonde catching your eye across the bar, you can't trust your body.

But your enemy can't trust his either.

A MEMORY TEST

Read the following list aloud to your test subject: Sugar. Cake. Candy. Cookies. Saccharine. Honey. Maple syrup. Pastry. Pie.

Now, without their noticing, hand them a different list, one that contains the same nine words you just read aloud, plus the word "sweet."

Explain to your victim that the words on the paper are the same ones you just read but that now they are out of order and, to test his or her memory, you'd like him or her to write the numbers 1 through 10 beside the words in the order that he or she remembers you reading them.

Most test subjects will not notice you've added another word because, in their mind, all the words are already associated with "sweet" to begin with.

We often hear of "repressed memory," where individuals are so severely traumatized by an event they "forget" the event ever happened. However, just as insidious, and potentially more dangerous, is the proclivity of human beings to fill in the blanks, to add "details" to incomplete and/or uncomfortable memories, to make those memories more compatible with their current view of the world. This helps the person avoid what psychologists call cognitive dissonance—mental anxiety created in the mind when we struggle to reconcile two opposing ideas or sets of facts.

In addition to being "deleted" (i.e., repressed) and augmented (i.e., added to), completely never-happened-in-a-million-years false memories can be all too easily manufactured.

Sometimes individuals manufacture false memories for themselves— filling in those bothersome blanks. Other times, false memories are "implanted" by outside influences—con men, cults, manipulators of various ilk.

Experiments have been done, to varying degrees of success, using hypnosis (sometimes in conjunction with drugs) to create false memories. Governments—including our own—did a lot of experimenting on this in the 1950s and '60s . . . and beyond? (See my book, *Mind Manipulation*, 2002.)

Cults and fanatical religious fundamentalists are notorious for planting false memories of parental molestation and even "Satanic abuse" in the minds of impressionable converts. (See Lung & Prowant, 2001.)

The malleability of memory is by now well documented, so much so that psychologists are finding increasing uses for deliberately reshaping a person's memories—from freeing patients from real memories of childhood abuse and trauma, to helping patients lose weight by implanting false memories that the patient hates certain high-calorie foods! (See Skloot, 2006:30.)

Says Skloot, "It is likely that false memories can be implanted only in people who are unaware of the mental manipulation" (p. 30).

So you can't trust your memory . . . but your enemy can't trust his either. Try to remember this.

A HEARING TEST

Ask your victim: "What's this spell: H-O-P." Their response: Hop. "P-O-P?" Pop. "T-O-P?" Top. "What do you do when you come to a green light?" Their response: Stop. Wrong, you "go" at a *green* light!

It's a law of physics that "objects in motion tend to stay in motion." This same rule applies to the Black Science. Humans are slaves to routine and repetition. We follow patterns, most often patterns that we are unaware of.

Make yourself aware of your own patterns . . . before your enemy does.

A READING TEST

Have your guinea pig (i.e., "victim") read this: GODISNOWHERE.

Did they read, "God is NOW here"? Or, instead, "God is NO where"?

Both answers (perceptions) are right . . . or are they both wrong?

"Relativism" is the philosophy that there is no absolute truth, that all seeming "truths" are relative, i.e., they are dependent on where you happen to be standing literally (physically) and figuratively (mentally). In other words, the truth depends on your perspective.

Cult leaders absolutely love "relativism," since it allows them to first make recruits doubt what they think is true about life, thus paving the way for what the cult leader decides is "reality."

Advertisers and politicians use this same sort of relativism when it comes to slinging around statistics.

Someone once said, "If you torture numbers long enough they'll confess to anything!" Still, you'd like to think (hope!) that something as straightforward as the "truth" of numbers couldn't be twisted and skewed. After all, 2 + 2 is always 4, right?

Well, maybe not always. Says Darrell Huff in his classic *How to Lie with Statistics* (1954):

> The secret language of statistics, so appealing in a fact minded culture, is employed to sensationalize, inflate, confuse, and oversimplify. . . . There is terror in numbers. . . . If you can't prove what you want to prove, demonstrate something else and pretend that they are the same thing. In the daze that follows the collision of statistics with the human mind, hardly anyone will notice the difference. (p. 8)

Such relativism goes by other names as well, "utilitarianism" and "egoism," for example. What's good for one's survival and personal happiness is therefore "moral." You can't trust statistics . . . or any weasely bean counter who's overly attached to them.

A COUPLE OF TESTS FOR THE EYES

What do you think this is?

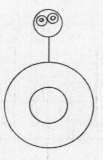

Figure 1.

7

Give up? It's a Mexican frying eggs 'n' bacon for breakfast! How about this one?

Figure 2.

It's a ten-foot length of common garden hose.

Just a couple light-hearted examples of how things can appear differently depending on your perspective.

Trust me, there are all too many more serious examples of how our eyes deceive us, how even our own *eye*-witness testimony may be suspect, especially when our perceptions have been filtered through the drama and trauma of seeing—let alone being the victim of!—a violent crime.

THINKING OUTSIDE THE BOX

How many squares can you count in this illustration?

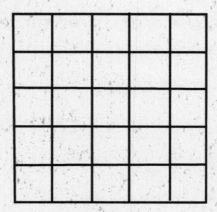

Figure 3.

Did you count twenty . . . or twenty-one . . . what about twenty-five?

It all depends on how you count them: the twenty obvious squares inside the larger square (Oops! Did you miss that one, number 21?). Then there are the four squares at each corner (made up of four smaller squares). Yep, twenty-five . . . unless you want to slide over one row from the sides to count even more squares (again composed of four smaller squares each).

Of course, it doesn't really matter how many squares you see . . . so long as you train yourself to *really look* and, more important, to look—think! — outside the box.

"Thinking outside the box" has become an overused catchphrase meaning we practice "nonlinear thinking."

Linear thinking harkens back to that predictable, routine thinking and behavior you've already been warned about.

Animals walk down the same woodland paths every day, stopping at the same watering holes every day. That's how an observant hunter knows where to set his traps, where to wait in ambush.

Let's assume that your enemy is, at the very least, an observant hunter.

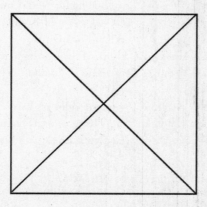

Figure 4.

In case you're wondering, it's a pyramid—top view.

But the question is: How many sides does this pyramid have? Take your time . . .

9

Four sides? Usual observation. But what about the base? If we counted the base, that would bring the number of sides up to five, right?

How about this: In addition to its usual four (or is it five?) sides, every pyramid also has an *in*side and *out*side. Total: seven.

Inside and outside? Just "semantics," just a play on words? Of course, but how many people have died simply because the wrong word was dropped in the right ear at the right time?

Words are never harmless. What is offensive to one may not be offensive to another.

That harmless "dumb blond" joke you tell at work today might have you on the unemployment line tomorrow.

What if your boss kept hearing rumors and then complaints that you were telling racist jokes? Suddenly one of his best workers (you) becomes a liability—literally, a legal liability.

Some people can't take a joke. In the same way some people—perhaps most people—can't think outside the box.

Many people are "concrete" thinkers, that is they take things literally. "A rolling stone gathers no moss" means to them, literally, that somewhere there is a big-ass rock rolling down a hill.

An "abstract" thinker, on the other hand, would understand this as a metaphor, meaning it's hard to put down roots if you're always traveling from place to place.

In fact, one of the most basic of psychology tests is called the Metaphor Test and is meant to determine if a patient is capable of abstract reasoning.*

As we shall see in chapter 2, concrete thinkers have trouble progressing from the appreciation (understanding) stage of a problem or strategy to the application stage, i.e., actually applying such strategies to deal with real world problems.

So who can you trust? At first glance it all looks pretty dismal.

But the good news remains that, despite protestations to the contrary,

---

*According to psychologist Jean Piaget, most children develop abstract thinking around the age of nine, although, for various reasons, ranging from congenital defects to stunted development, some adults have a "retarded" degree of abstract thought. Such individuals take everything literally.

when it comes to the Black Science, it's a level playing field. Your enemy has the same weaknesses you have.

Even more encouraging is the fact that the more aware you are of the weaknesses you possess, the better your chances of overcoming those weaknesses . . . or at least learning to hide your weaknesses better from those who would use it against you.

Study on!

2

Appreciation vs. Application

"Do the thing, and you have the power."
—Nietzsche

Admiring Michelangelo Merisi Caravaggio's paintings, appreciating the way this Baroque genius mastered the delicate play of chiaroscuro, doesn't mean you can paint like Caravaggio.

Likewise, watching Bruce Lee movies, no matter how much you appreciate his style, won't make you a martial artist capable of applying Bruce's moves when Death decides to take the same shortcut through that dark alley as you.

There's a big difference between "appreciation" and "application," no matter the field of endeavor.

For example, any serious student of the Black Science knows that *The Art of War* by Sun Tzu *is* required reading . . . over, and over, and over again!

And, while it's easy to "appreciate" Sun Tzu's undisputed mastery of strategy, novices often have trouble applying the subtleties of Sun Tzu's *Ping-Fa* to their everyday lives.

Appreciation is easy, application is not. What's sweet in the mouth may be all too bitter in the belly.

WHAT WOULD SUN TZU DO?

Sun Tzu warns us never to fight an enemy whose back is to a mountain or to the sea (since trapped men—with no seeming way of escape—have no choice but to fight to the death).

12

The Master goes on to admonish us to always leave the enemy with a way out. (Of course, you can always have an ambush set up farther down the road, but that's a strategy for another day.)

So what is the average Joe Blow to make of this advice from the Master? The average Joe is not commanding any Chinese armies and, while he may "appreciate" (understand) Sun Tzu's strategy as it applies to the maneuvering of thousands of Chinese troops 2,500 years ago . . . what's that got to do with Joe's boring life today, in the here and now?

This is what the youngsters like to call "keepin' it real."

So how about this scenario:

Average Joe's waiting in a checkout line when a stranger bumps into him. "My bad," the stranger apologizes, using the current slang.

"Why don't you watch where the hell you're going!" Joe screams at the stranger. "What the hell's your major malfunction, numbnuts!"

"Hey, I said I was sorry," the stranger protests, his own temper starting to rise.

"Well 'sorry' didn't do it, you did. What are you, on drugs or something?" Joe rants.

"Hey, screw you, buddy!" the stranger explodes, as he and Average Joe square off. . . .

Or, remembering Sun Tzu's advice not to (figuratively and literally) back an enemy into a wall, Joe accepts the stranger's apology and allows him an "honorable" way out of the situation.

How many arguments escalate simply because one party to a disagreement backs the other against a wall, failing to give them a way out.

Savvy business negotiators (Black Science graduates all!) know the value of always allowing disputing parties an "honorable" way out.

Hostage negotiators are taught that, when lives hang in the balance, it's a good idea to show the crazed hostage taker that he's *not* backed himself against a wall, that there's an honorable way out—one that doesn't involve body bags.

On the other hand, unscrupulous con men purposely trap their victim in no-win situations, with seemingly no escape . . . before then offering them—what seems to be!—a reasonable, face-saving way out.

So it's not enough to only appreciate the strategies of the great movers and shakers of history, the righteous as well as the ruthless, the key is learning the application of those strategies, tactics, and techniques to our time and place. This takes a little practice in abstract thinking.

WHAT WOULD MUSASHI DO?

In Miyamoto Musashi's *A Book of Five Rings* we are taught that when unable to defeat an enemy by attacking him head-on, we should adopt the strategy of "cutting at the edges."

No, this doesn't mean we actually have to attack our enemy with three-feet of Japanese steel. What Musashi is telling us is that, when faced with a difficult, seemingly impossible situation or foe, knowing that a direct confrontation will not work to our advantage, we must attack from an oblique angle.

Say that, rather than actually trying to imbed our razor-edged katana into our enemy's skull, what we're trying to do is plant a suggestion (true or false) into his mind.

Now whereas he wouldn't give us the time of day, we might try obliquely sending our message via one of his friends, relatives, coworkers, or simply the guy he buys his newspaper from every morning.

Perhaps his wife is at the Laundromat or beauty parlor when she "accidentally" overhears someone spilling the beans about a piece of information that might interest or otherwise affect her husband. Whereas her husband might not accept the information coming directly from you (a competitor, or outright enemy), when his wife passes it on to him, there's a better chance he will give it a second look and a third thought.

Advertisers often use indirect "cutting at the edges" ploys by hiring "product placement actors" to frequent a business and ask if they stock a new product (one they know the store doesn't carry), making it more likely the owner will then buy the product when the salesman for that particular product just happens by the next day.

Internet advertisers use this same ploy, hiring workers to deliberately haunt chat rooms, stirring up conversations designed to intrigue others enough to log on to a specific site.

Back in the day these were called "agents provocateur."

So how do you progress from mere "appreciation" of the Black Science to successful—and profitable—"application" of its strategies?

Like everything else in life: You practice. Starting now:

WHAT WOULD GELON OF GELA DO?

Early in the sixth century BCE, the Greeks and Carthage (the same North African city-state that would later give the world Hannibal and his elephants) were recklessly eyeballin' one another over the economically strategic island of Sicily.

In 733 BCE Greek traders from Corinth established the port of Syracuse on the east coast of Sicily. Syracuse quickly became the prosperous template for other trading cities on Sicily.

Soon after, in 583, traders from Crete and Rhodes founded the city of Gela on Sicily's southern coast. About a century later, similar seafaring traders founded Akragas farther west down the coast.

Thus, by the beginning of the fifth century, Sicily was already sharply divided into rival spheres of influence: Carthaginians in the extreme west, and Greeks controlling the rest of the island.

The most ambitious (read: "dangerous"!) of these Sicilian Greeks was Gelon, ruler of Gela.

"Ambitious, dynamic, anti-Carthaginian imperialist" (Soren, 1990), Gelon dreamed of a unified Greek Sicily—albeit one under his control. To accomplish this, Gelon first formed an alliance with the equally ambitious King Theron of Akragas, and began systematically bringing smaller cities in the area under Gela's influence.

Then Gelon waited, biding his time. . . . Of course, that doesn't mean he wasn't active.

Knowing how to sharpen your sword is just as important as knowing how to wield it.

Besides, Gelon didn't have all that long to wait because—surprise!—there soon arose an internal revolt in Syracuse spearheaded by its landed nobles; a revolt—surprise number two—instigated and fanned by Gelon's vast network of agents.

Eventually, after the usual internecine slaughter, the warring parties decided an outside mediator was needed to help sort out the continuing dispute and disorder.

Ready for another surprise? Just out of the blue, someone suggested Gelon of Gela since he was renowned (thanks to his successful propaganda campaign) as "a true man of the people." (Recall that Hitler was also a "compromise candidate.")

After the usual false humility of pretending to turn down the initial offer (often a valuable ploy to get someone to sweeten the pot), Gelon "reluctantly" took power in Syracuse without firing a shot, promising to relinquish power as soon as the crisis passed. (Like we haven't heard that one before!)

No sooner had Gelon taken power than he put the next phase of his master plan into operation.

In short order he had moved most of the people of Gela, as well as people from other Gelon dependencies into Syracuse.

Overnight, the long-time citizens of Syracuse became a minority in their own city. And "Gelon of Gela" had suddenly become "Gelon of Syracuse," ruler of all of southeastern Sicily. According to Soren (1990):

> It was a dangerous moment. Like Hitler in Europe, the ruthless Gelon was making alliances of convenience, displacing peoples and gobbling up land. (p. 55)

All was looking good for Gelon except that he forgot the adage, old even back then: Better a true enemy than a false friend.

Busy consolidating his power in Syracuse, Gelon failed to keep an eye on his ally, Theron, who decided to do a little expanding of his own by gobbling up Himera, a prosperous port city on the north coast of Sicily.

Escaping the capture of his city, Terillos, king of Himera, immediately pleaded with his allies, including Carthage, for help to regain his crown. This was just the excuse Carthage had been waiting for. Carthage declared war on Theron and, by proxy, his ally Gelon.

In short order Carthage took Himera back from Theron . . . and stayed. Himera became Carthage's beachhead and staging area on Sicily from where they could build up their forces.

A "crafty and brilliant cavalry leader with wide experience" and, even more important "a master of dirty tricks espionage" (Soren, 1990: 55), Gelon was smart enough to realize when his own empire was being threatened by Carthaginian hegemony. He also realized that no one could go oar to oar against the Carthaginian fleet on the open sea. So he decided the best course of action was to force them to fight on land, where the seagoing Carthaginians were literally fish out of water.

It is said, "Chance favors the prepared mind" and never was this truer than when Gelon's vast network of spies intercepted a secret Carthaginian message ordering their troops to assemble at Selinas, west of Himera Harbor where the Carthaginian fleet was anchored.

Quickly disguising his own men as Carthaginians, using the secret message to grant them safe passage, Gelon staged a sneak attack west of Himera designed to draw off the Carthaginian forces.

The plan succeeded perfectly. As the Carthaginians rode west to counterattack, Gelon's main force attacked Himera, burning all the ships in the harbor.

So devastating was Gelon's attack that Hamilcar, the Carthaginian general in charge of the fleet, committed suicide by jumping onto a sacrificial fire, ending Carthage's claims to Sicily . . . at least for the time being.

With Carthage out of the picture, Gelon of Syracuse prospered.

So what's all this got to do with me, you ask?

Let's see what lessons we can glean from Gelon of Gela, Syracuse, and all points Sicilian that compare with, and might apply to, our contemporary world (see pages 15–16).

GELON of GELA
Job description: Fifth century BCE tyrant

I.   Forms alliances with already powerful King Theron of Akragas. "The enemy of my enemy is my friend."

II.   Practices patience, biding his time. Goes out of his way to appear nonthreatening, unambitious, all the while expanding his intelligence gathering network, infiltrating his spies and agents provocateur into strategic positions in rival ports and courts.

III.   Civil war breaks out in Syracuse—civil war Gelon's agents helped instigate. On his orders, his agents continue to instigate further unrest among the landed nobles of Syracuse while simultaneously whispering into their ears the need for an "outside arbitrator" . . . someone like Gelon!

Thanks in large part to Gelon's machinations, the situation in Syracuse finally reaches crisis point and a coalition of Syracuse political parties BEG Gelon to come in as "Tyrant" (i.e., temporary ruler with wide-ranging powers to quell disorder and settle disputes).

IV.   Once firmly in place, free to exercise power, Gelon floods Syracuse with immigrants from his native Gela and his other dominions until he is surrounded by those fiercely loyal to him. These newcomers quickly come to dominate the now minority citizens of Syracuse.

V.   Entrenched as Tyrant, surrounded by those loyal to him (and beholden to him for their new-found wealth and power they've usurped from the Syracuse natives), Gelon consolidates his power by purging anyone who formerly opposed him and/or might conceivably pose a threat to him in the future.

VI.   From his new stronghold in Syracuse, Gelon continues to expand his empire.

Figure 5.

JOE of BLOW
Job description: Modern-day aspiring executive

I. Joe gets to know fellow office workers (especially their hidden peeves & grievances); secretaries; other bosses; that office boy who delivers the mail, who has ambitions of his own; even the cleaning lady who can give you your rival's trash to sift through for blackmail material—anyone who might let leak valuable insight and information. He perfects the art of ass-kissing. (It's not "ass-kissing" when you *know* you're ass-kissing. Then it's called "strategy"!)

II. Joe bides his time. He doesn't want to appear too eager, or a noticeable threat to already established bosses and/or other aspiring "Joe Blows" (i.e., the "competition"). His spies and informants in place, he begins collecting data on office politics, noting where new alliances might be formed, where existing "old Boy Network" alliances might be sabotaged.

III. A crisis occurs, ore Joe saw coming (more than likely because Joe *created* it!). Just when all looks lost, one of Joe's secret agents suggests to the bosses that Joe might be able to solve the problem because he or she remembers hearing about a similar problem Joe solved at his previous job.

Against his protestations ("false humility"), Joe is "drafted" to solve the problem, a problem guaranteed to be above Joe's current level of job description.

IV. Joe solves the problem in record time (but making sure everyone sees how much sweat and effort, dedication & hard work he put into accomplishing the goal). Joe is rewarded by being promoted to the level of the problem. (The boss's thinking being, "We need to keep this handy fellow where he can keep an eye cut for that kind of thing, to make sure it doesn't happen again!")

V. Quickly—quietly!—entrenching himself in his new position, Joe brings in his own people—promoting those who helped him rise to the top. These Rewarded Ones will now be fiercely loyal to Joe, guarding his back from other "Joe Blows" still stranded below.

VI. Secure in his power and position, a picture of Gelon of Gela hanging in his new corner office, Joe schemes to expand his "empire," eyeing new conquests using his tried-n-true method.

Figure 5 *continued.*

This side-by-side comparison makes it a little easier to understand—appreciate—how the techniques used to obtain position, prestige, and ultimately power haven't really changed all that much down through the centuries. That's because human beings haven't changed all that much down through the centuries.

You read ancient books of strategy like Sun Tzu's *Ping-Fa*, written 2,500 years ago on the other side of the world, a treatise originally scripted by and for Chinese warriors, and while you may "appreciate" Master Sun's exploits and insights, you still stumble when it comes to finding contemporary comparisons and practical applications for such battlefield wisdom.

But if you really take the time to study his *Ping-Fa*, you'll realize Master Sun spent twice as much time trying to discover what an enemy had hidden in his heart, as opposed to what that enemy had parading around openly on the battlefield.

Sun Tzu knew, and now you do, too, the key is to look at the people. People haven't changed since Sun Tzu's time, nor since Gelon's time, nor since Machiavelli's.

And you have only to read Shakespeare to realize that, 400 years after the Bard's time, people are still stressed out and manipulated by love (*Romeo and Juliet*), jealousy (*Othello*), betrayal (*Macbeth*), and revenge (*Hamlet*).

Technology changes. Sensibilities and morals *appear* to have changed down through the centuries. But, at the core, peoples' lusts and fears and petty hatreds have remained intact. Today we just hate more in private and screw more in public.

Some 2,500 years ago Sun Tzu told us if we want to knock a rival general off his square we should tempt and distract him with jade (riches), the lure of sensual pleasures, and anger.

Hundreds of years ago Miyamoto Musashi pointed out that if we yawn, we can cause an opponent to involuntarily yawn—causing him to drop his guard just long enough for us to make a telling thrust.

Are people really any different today? Are we not still swayed by promises of sensual delight, by rewards of riches? Are we not all too easily distracted and galvanized to violence by anger?

Any time you begin to doubt the insights of these ancient strategists, just try yawning in a room full of people. And then tip your hat to Musashi.

Ancient Chinese and Japanese strategists always looked for what they called the "Five Movers" or the "Five Weaknesses" that all people possess to some degree or another.

In Black Science today we call these five weaknesses the "Five Warning F.L.A.G.S.": Fear, Lust, Anger, Greed, and Sympathy; any one of which, when we are unaware of them, can undermine our relationships and/or our business dealings.

Of course, once you are aware of these potential weaknesses, weaknesses all people possess, we can weed them out of our own mental garden . . . while liberally strewing seditious seeds in our enemy's backyard.

Bottom line: The secret for progressing from passive appreciation to active application of any particular strategy, tactic, or techniques of a Sun Tzu, a Machiavelli, a Hitler, or a Donald Trump for that matter, is first and foremost *self-awareness*. (More on this in chapter 6, Mind Control by the Numbers.)

As long as we can use such strategies to (1) better guard what's ours and (2) turn such strategies to our advantage, it matters not whether those strategies were birthed long ago, or were born yesterday, in Peking or Peoria.

For, while "East may be East, and West may be West, and never the twain shall meet," P. T. Barnum has yet to be proved wrong.

Part II

EAST IS EAST . . .

"Average Easterners and average Westerners, who come into accidental contact, no doubt find themselves 'poles apart,' as the English saying goes. Their habits of life are not the same; they appear to have different habits of thought, and a different outlook on the world. Consequently their impressions of each other are, when untinged by fanatical race prejudice, always interesting, sometimes novel, and perhaps, amusingly odd; but they are rarely, be it borne in mind, based upon intimate knowledge or genuine sympathy in the wide sense of the term."
—Nawab Lada of Sardhana, in Shah, 1938:351

"The East is East and the West is West, and never the twain shall meet," declared Rudyard Kipling (1865–1936), and while speaking specifically of the Indian subcontinent he knew and wrote of so well, it is not so bold an extrapolation for us to apply his observation to Asia as a whole.

Since the return of Marco Polo (1254–1325) the East has always fascinated the West and, should they care to admit it, we might find those in the East just as perplexed—and perhaps intrigued—by Western ways.

Kipling himself, Hermann Hesse (1877–1962), and a host of others wrote on the mystery and intrigue of the Far East. Hesse, along with such intrepid cultural explorers as Theosophy's Helena Blavatsky and Tibet's Evans-Wentz, fearlessly delved deep into the mysteries of the East, finding Asian philosophies and techniques famed for first focusing the mind, in preparation for opening it to a broader understanding and an ultimate unleashing of its latent powers.

But, no matter the tool, in the wrong hands it can all too easily become a weapon. Of course, the opposite is also true, whether speaking of the resilience of plow-steel beaten into sword blade or the keenness of mind and the sharpness of tongue.

British authorities took the treacherous techniques of intrigue, infiltration, and strangulation from the dreaded Indian Thuggee killer cult and turned these murderous methods to good use, training their own British commandos (Lung, 1995, 1998).

And hasn't the physical and philosophical disciplines of a myriad of Asian martial arts enriched—and perhaps saved—the lives of countless Westerners?

Thus, good and evil are ever in the eye—and fist!—of the beholder. Only the hand that holds the blade has cause for shame, never the blade itself.

Still, for every selfless Gunga Din who will befriend us during our journey to the East, we must beware running afoul of an insidious Dr. Fu Manchu. For every Gandhian pacifist who offers us a light to guide us on our way, we must remain alert to being stalked by Shinobi shadow-warriors. The brighter the light, the darker the shadow.

For every Alan Watts and Baba Ram Das bringing us back Zen and yoga enlightenment, we get an Adolf Hitler all too eager to master the "dark side" of eastern occultic mind manipulation in order to further plunge the West—and the world!—into darkness: "Although they sound like the fantasies of a madman to most of us, [Hitler's] ideas were derived from Tibetan magic. So was the swastika. Many Nazis were initiated into secret occult societies all over Germany and trained in Eastern black magic" (Ostrander & Schroeder, 1970:327).*

But sometimes we see more in the darkness than we do in the light.

As the twentieth century waxed and waned, this relationship between East and West became increasingly symbiotic, often to the benefit of both—sometimes to the detriment of all.

For the West, it can be argued, has had just as much influence, overt and

---

*For more on this see information on Nazi intercourse with Asian secret societies including the Green Dragon Society and Japan's infamous Black Dragons in *Mind Manipulation*, (Lung & Prowant, 2002) and *Ninja Craft* (Lung, 1997)

covert, on the Eastern mind. And not just forcing the Japanese to trade in their Samurai swords for Louisville sluggers.

Didn't Freemasons and International Finders help found the Vietnamese religion of Cao Dai? (See Lung, 2002; Lung, 2006.)

But then again, at least one Western researcher has credited the Buddha with inspiring Freemasonry, of being "the first Masonic legislator," while tracing Masonic ritual back to Buddhist monks (cf. Robertson, 1991: 184–185).

And there are increasing revelations coming to light of Asian techniques of mind manipulation and mind control being experimented with by shady Western individuals, cabals, and even governments (cf. Russell, 1992).*

We'll delve into this Western "mis-use" of time-honored Asian techniques of meditation and mind expansion in subsequent chapters.

The Asian culture overall, as well as various countries within that culture, not to mention secretive and self-serving cults and cadre within those countries, have all developed numerous and distinct schools of philosophy and psychology, any one of which alone, and all collectively, can give us unique insight into the maintenance and manipulation of the human mind.

Predictably, down through the ages, there has been and continues to be considerable interchange—read: skulduggery, theft!—between Asian countries and cadre when it comes to getting the upperhand in the battle for the minds of their fellows. Of course, we in the West are just as guilty.

Not to fret: the brightest of diamonds are often found in the muddiest of holes.

Remember that to the utilitarian Black Science student, it matters not the source of a scheme or skill, so long as it proves itself in the end.

And since Sun Tzu warns us that "victory has never been associated with long delay," endeavoring to be expeditious, we will divide our study of Asian Black Science into three basic "spheres of influences": (1) East Indian, (2) Chinese, and (3) Japanese.

---

*Do not confuse the terms "mind manipulation" and "mind control"; they are not synonyms. As used at the Black Science Institute, "mind manipulation" is more generic, referring to any and all attempts to influence the human mind, most often for dubious and diabolical intent. Mind manipulation ranges from Madison Avenue advertising to cult conversion. "Mind control," on the other hand, refers specifically to the goal and end product of deliberately malicious mind manipulation, i.e., total mental (thoughts and emotional) control over a victim.

And as we study these three spheres, each with their own unique ways of delving into, deciphering, and, when deemed necessary, dominating and/or destroying the human mind, let us always look for similarities, rather than be distracted and dazzled by seeming differences.

Whether, in the end, we will decide that there are indeed major differences-unbridgeable gaps——between the way Easterners perceive the world and the way we in the West see it, remains to be seen.

3

India: The Mother
of All Mothers

*"If I want to understand somebody, I cannot condemn him,
I must observe, study him. I must love the very thing I am
studying. If you want to understand a child, you must love and
not condemn him. You must play with him, watch his
movements, his idiosyncrasies, his ways of behavior;
but if you merely condemn, resist or blame him, there is no
comprehension of the child."*
—J. Krishnamurti, 1954:42

India has been called the "Mother of Nations," a richly deserved title.

For much of the known history of Asia, India has been the creator and catalyst for enlightening cultural movements and specialized cadre of unique thinkers—from conquerors to cultists, goddess-inspired killers, to Gandhi. For always within India there has been a fascination for deciphering and disciplining the mind. For the mind is the gatekeeper to the realms of the spirit.

As a result, so many fascinating, and frightening, schools of thought have come out of India.

Yes, it is true India gave birth to the merciless Thuggee cult of stranglers (Lung, 1995). But India also gave us the Buddha. (See Nine Hounds, later in this chapter.)

India is acknowledged as the maternal hearth of the Asian martial arts—

these Indian martial arts then leading to Chinese Kung-fu, Japanese karate, and a hundred others styles of self-defense. Of course these arts in turn spawned deadly dark knights like China's Moshuh Nanren and Japan's Shinobi Ninja.

Indian yogis and fakirs are renowned for their stoic asceticism. Yet from this same school of discipline came the notorious Tantra sex sects and erotic writings such as the *Kama Sutra* (Lung, 2002).

Dozens of strains of Indian yoga taught Indian seekers—and eventually the world—how to cleanse and calm the bothersome human body, and how to purposely circulate the mysterious inner kundalini force that animates the body, in preparation for meditation and enlightenment.*

Of course, control of one's own body and mind all too often proved too much of a temptation for some unscrupulous yogis, and so they were soon using their newfound powers (Skt. siddhas) to dominate those less adept, those less suspecting.

At first mastery of *Kundalini-Chi* flow led to the development of the healing arts of Chinese acupuncture and Japanese *shiatsu*. Only later, perhaps inevitably, would come *Dim Mak*—the dreaded "death touch" capable of killing an enemy instantly, without leaving a mark, simply by interrupting an enemy's chi flow (Omar, 1989; Lung, 1997; Lung, 2002: "The One-Eyed Snake").

Life is ever a two-edged sword. . . . Death, the scabbard that finally sheaths it.

Thus, for every bright and shining son given us by Mother India . . . it seems she has brought into this world just as many bastards—deformed and diabolical children!

No matter their lineage or legitimacy, there are lessons to be learned from both.

> *"Most of us are not creative; we are repetitive machines, mere gramophone records playing over and over again certain songs of experience, certain conclusions and memories, either our own or those of another."*
> —J. Krishnamurti, 1954

---

*Kundalini is akin to the Sanskrit "prana" (spiritual breath) which is called chi (qi) in China and ki (pronounced "key") in Japan.

SIX SENSES, FIVE VIRGINS

*"They don't know the day / is the dark's face / and the dark
the day's."*
—Prabhu Allama, Vacana poet

The great tree of Indian Yoga (Skt. "yoke") has many branches that have, down through the centuries, borne more than one picking of queer and questionable fruit.

When most Westerners think of yoga, they think of Hatha Yoga, with its impossible-to-contort-into postures (asanas). However, there are many other schools of yoga; some yogas concentrate on ascetic practices (your basic sleeping on a bed-o-nails), the observation of traditional rituals (Raja Yoga), while others concentrate on selfless acts and devotional duties (Bhakti Yoga).

Kundalini Yoga specializes in techniques designed to stimulate and direct the Kundalini force within the body.

Further outside the mainstream is Tantric Yoga. Over the years, in large part due to the prurient interest of Western reporters, Tantric Yoga (which boasts both a Hindu branch and a Buddhist branch) has gotten a bad reputation for being licentious and perverted for allowing its members to eat meat and drink alcohol (both forbidden in both orthodox Hinduism and Buddhism), and for giving its blessing to a form of two-person meditation that prolongs, and heightens, sexual pleasure—often delaying orgasm for hours! All in the name of enlightenment, of course.

This is hard for Westerners to understand but it goes something like this:

In yoga there are two basic paths: the right-hand path and the left-hand path.

Those yogis following the right-hand path (the more "traditional" of the two) abstain from worldly pleasures, including meat, alcohol, drugs, and sensual congress. By doing so it is believed that the flame of worldly passion will die out for lack of fresh fuel. Once this flame of passion dies out, the yogi—no longer distracted by worldly temptations—becomes enlightened.

The left-hand path—uh, on the other hand—is of the opinion that there's nothing wrong with seekers after enlightenment indulging themselves.

They point out that, since God is "every, always, and perfect," there is not a little more of God "here" and a little less of God "over there." Therefore, they contend, a seeker is just as likely to "find God" in a whorehouse as in a temple. Therefore meat and alcohol and all manner of sex, drugs, and rock 'n' roll are permitted to a practitioner of the left-hand path.

Left-hand path yogis reason that, again, passion is like a flame, but a flame they choose to fan by indulging. By indulging their passions, they eventually become "burned out," i.e., they exhaust all the fuel of passion and, on its own, the flame of passion sputters and dies—and the yogi attains enlightenment.

Not surprising, many Westerners find the study and practice of Tantric Yoga much more suited to their sensibilities!

Whichever branch of the yogic tree we decide to study, all such studies begin with an understanding of the "six senses," for it is a yogis' mastery of the six senses that leads him to the "Five Virgins."

For yogis, there are the universally accepted five senses of smell, taste, sight, touch, and hearing, to which they further add "heart" or "feeling." (See page 31.)

THE SIX SENSES

Element	Earth	Water	Fire	Air	Sky	Space
Outer Sense	Smell	Taste	Sight	Touch	Hearing	Feeling (heart)
Inner Senses	Citta	Buddhi	Ahankara	Manas	Jnana	Bhava
Sakti type (Divine power)	Kriya (action)	Jnana (knowing)	Iccha (will)	Adi (creativity)	Para (power)	Cit (intelligence)

The Five Virgins

Kriya   (The Power of Action): The power to instantly change thought into action; energy (thought) into action. $E = Mc^2$

Jnana   (aka *Mantra*, the Power of the Word): The power to instantly create effect with just a spoken word; to control self and others through vibrations (e.g., words).

Iccha   (The Power of Wish and Will): The power to simply desire an object and/or outcome and thus bring it into reality.

Adi   (The Power of Primal Creativity): The power to innovate and create completely new ideas. The power to create "something" from "nothing."

Para (Transcendent Power): The power to create effect(s) that seem outside the laws of physics; to create "magical" effect; to transcend physical and mental restrictions of thought and action.

Mastery of the initial five senses unlocks "powers" (Skt. siddhas) that allow yogis to free themselves from the physical and mental restrictions that hold the rest of us down.

Keep in mind that, from a Black Science perspective, our enemies are all susceptible to these five senses and, as we will discuss in a minute, once we have mastered our own senses, we can all too easily use the fact that our enemy has not mastered his senses to our advantage.

Thus we have two reasons for mastering our five senses: First, in order to better regulate our own lives, in order to better protect ourselves from the potential disaster of allowing our senses to run wild, ruling us and not the other way around. Second, according to yogic teaching, mastery of our five senses gives us power to create effects that affect—manipulate—the world around us, and the people in that world.

On a purely physical level, learning to control our five "animal" senses takes us to a new level of awareness. Practicing this "advanced" level of awareness often makes it appear to others that we possess some sort of ESP when, in fact, all we are doing is (1) not allowing ourselves to be distracted by the itchin' and bitchin' of our five senses, freeing us up to (2) simply pay closer attention to the world around us.

With the mastery of each of the five senses, we are rewarded by a gift of one of the "Five Virgins," so-called because these are considered to be "pure" gifts from a divine source.*

Mastery of Kriya

Kriya—the sense of smell—gives us the power to act instantly, to turn our thoughts into action. How often have you been working or perhaps walking down the street, when a pleasant smell instantly made you forget your immediate task? How often has a smell suddenly made you nostalgic,

---

*In Hinduism this divine source is called Sakti, the feminine consort and counterpart of a god, in this case, Shiva, the Destroyer.

instantly transporting you from the present into some dreamy, idyllic pleasurable past experience? It happens all the time, testifying to the power of smell to instantly change our mind, to transform thought into action.

Animals use pheromones to communicate, to attract mates, in some instances even to protect themselves. Consider the skunk.

Having lost touch with our bestial roots, humans don't like to be reminded how often they are literally led around by the nose. The good news, however, is our enemies all suffer from this same failing.

Mastery of Jnana

Jnana gives us the power to create—and destroy—with just a word. Physics 101: Remember that all things are simply collections of vibrations/vibrating at various wavelengths. A brick wall is of a "denser" wavelength than your body, which is why your body can't walk through a wall. Water, on the other hand, has a higher vibration rate (is less dense) than your body so your body can pass through water.

The vibrations of one thing often influence another—the classic example being the opera singer whose voice shatters crystal—a more plausible explanation given for the sound (vibrations) of Joshua's horns felling the walls of Jericho.

Likewise, we are all aware of how certain spoken words (endearments, racial slurs, etc.) can affect us emotionally; some soothing us when we feel depressed, others inciting us to violence in an instant.

According to yogis, mastery of our sense of taste strengthens our power of speech. Literally, instead of our tongue ruling us (e.g., through hunger), we now rule the tongue.

Throughout Asia are found "mantras" (spoken power formulas, often a single syllable) that are believed to affect the body and mind—not only of self but of others.

Each of the Five Virgins has specific sounds (i.e., vibrations) associated with it (see figure 6).

Traditionally, these "mantras" are used to help yogis control the five senses and ultimately gain enlightenment.

As a natural by-product of leading an ascetic life and having a dedication to meditation, the innate physical and psychical powers within the yoga student unfold naturally, at a gradual pace, as the student's corresponding wisdom and ability to assimilate such power increases.

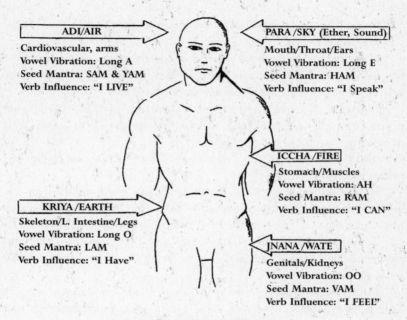

ADI/AIR
Cardiovascular, arms
Vowel Vibration: Long A
Seed Mantra: SAM & YAM
Verb Influence: "I LIVE"

PARA /SKY (Ether, Sound)
Mouth/Throat/Ears
Vowel Vibration: Long E
Seed Mantra: HAM
Verb Influence: "I Speak"

ICCHA /FIRE
Stomach/Muscles
Vowel Vibration: AH
Seed Mantra: RAM
Verb Influence: "I CAN"

KRIYA /EARTH
Skeleton/L. Intestine/Legs
Vowel Vibration: Long O
Seed Mantra: LAM
Verb Influence: "I Have"

JNANA /WATE
Genitals/Kidneys
Vowel Vibration: OO
Seed Mantra: VAM
Verb Influence: "I FEEL"

Figure 6. "Attack of the Five Virgins"

Of course, as is to be expected in all human undertakings, there are always those opportunistic and impatient few, those more rapine, not content to allow nature to lift her skirt in her own good time.

For beyond the promise of other-worldly enlightenment, there was—and still is—the temptation of temporal power for any seeker sly enough to seduce (or squeeze!) the Five Virgins into prematurely revealing their charms.

For example, Kriya, associated with the element Earth, controls the body's skeletal system, legs, and large intestine; pretty much the body's structural support system.

The corresponding psychological component to this is that Kriya acts as natural Prozac, controlling our mental equilibrium as well, our balance of thought and therefore action. This is why the "power" given us for mastering Kriya is the power to instantly change thought into action, energy into effect.

To "cleanse" our sense of smell and help and activate this power, Yogis often chant the Seed Mantra "LAM" and sometimes "O-LAM," since a long "O" sound is the major vowel vibration a person dominated by the element Earth responds to best.

It is easy then to see how, once we discover that our enemy is predisposed/dominated by the element Earth, to seed our speech and/or our writings aimed at him with words that emphasize the sound of long "O".*

You can also influence your enemy by peppering your spiel with his particular "verb influence," a phrase that best relates to—and captures—his personality.

In the case of Kriya, the verb phrase "I have" catches and holds the person's attention.

So how do you know what a particular person's verb influence is? You listen. That phrase most oft repeated by the person links them with a particular element, etc., allowing the Black Science adept to take it from there.

The person may not always repeat their actual verb influence, for example verbalizing "I have" over and over. Sometimes they use synonyms that mean the same thing.

Black Science mastery of the Jnana aspect of our five senses, at the very least, helps us learn to listen and "hear" better, giving us the advantage over our enemy by allowing him to give himself away.

What's that old saying? "A man begins by telling you what he knows but, if you let him talk long enough, he begins telling you what he doesn't know!"

Mastery of Iccha

Iccha hones our desire to a keen edge, giving us the power of "wish and will," the ability and skill to bring a desired object or event into reality.

The wisdom of Buddhism begins and ends with the Buddha's realization that we suffer because we either want things (objects and events) we can't reasonably obtain or else things we don't really need. Once we can overcome these irrational wants into "right desires," we are one step farther along on the road to Nirvana-enlightenment . . . at least according to the Buddha.

---

*Writing words implanted with specific vowel vibrations works just as well as actually speaking such words to him because, when we read words and phrases, our brain "pronounces" the words silently. So whether we hear these words (subliminal sounds) or see them in our mind, our brain still "hears" and internalizes them.

The Buddha spent much of his life mastering the ways of various yoga schools before ultimately attaining enlightenment, so it's not surprising that he should use the first two of his "Four Aryan Truths" to warn us of the danger of having "false desires."

It is also no mystery that Iccha is controlled by the element Fire, which in turn controls our sense of sight (since we are so often led by our sight when it comes to the things we desire).

How many times a day does television show us new products and gadgets that a minute ago we didn't know existed and now, after seeing them, we can't live without?

Likewise, don't we fall in love "at first sight"?

Human beings are very visual-oriented creatures. Seeing leads to desire (Fire) and our troubles begin.

Conversely, we learn to control desire, through willpower (Iccha) and in so doing strengthen our willpower. This process feeds on itself—like a "feedback loop."

The exercise of "will" lets us break free of desires, and that will—strengthened through use—now brings us our heart's desire—anything we can *wish* for we can now *will* into existence.

This is why the verb influence for Iccha is "I can," the will declaring itself.

Negatively, a person dominated by Fire isn't in control of his desires and so his assertion of "I can!" is only arrogant boasting.

Iccha/Fire pools in the stomach and muscles of the body.

Those predisposed/dominated by Fire are thus easily led astray by the "three A's":

- Their Anger (evidenced by muscle tension)
- Their Appetites (symbolized by the stomach)
- Their Arrogance ("I can!")

Discovering that an enemy is ruled by a Fire personality makes it easy to craft snares and dig pitfalls for him based on his predisposition toward anger (one of the Five Warning F.L.A.G.S.), appetites, and arrogance—personality flaws that, if he is like most people, he remains blissfully—dangerously!—ignorant of.

Mastery of Adi

Adi unleashes the power of primal creativity lying latent within us all; the power to innovate and invent, to see a problem from all angles before creating an appropriate and often unique solution to that problem.

This is the power used by that first caveman who rubbed two sticks together to get fire.

Adi is linked to the element Air, for it is from this ephemeral realm of pure ideas that our hopes and dreams and desires are drawn, to become manifest in the physical world through our focus and intent.

Once manifest we can now physically touch the objects and participate in the events we helped bring to fruition. Thus ADI is also associated with the sense of Touch. Therefore this is the sense we must conquer in order to "release" our inner creativity. We can do this by meditation on the mantra *"Ay-Sam-Yam."*

An enemy predisposed/dominated by the element Air will often have trouble with his heart and lungs, and can be a prime candidate for a heart attack or stroke. He may also suffer from asthma and high blood pressure. Turn up the "heat" and pressure in this guy's life and he probably won't be around to bother you for long!

The verb influence for Adi is the positive assertion "I live!", a simple celebration of life.

However, for a negative person dominated by this kind of mind-body attitude, "I live" becomes an assertion of selfishness.

Positive Adi individuals are balanced and loving, thus they are prone to find love in their lives.

Negative *Adi*-dominated people are out of balance and hate filled. Correspondingly, they draw negative and destructive influences to them, reinforcing their generally cynical outlook.

To bring down an enemy who has a cynical outlook, you have only to agree with his cynical outlook, reinforcing his cynicism, fanning the flames of his hate like a strong wind until it literally sucks the Air out of his body!

Mastery of Para

Para focuses our innate mind and body energies to the point where—to others—we seem to possess transcendent, almost magical, powers to affect the people and places around us.

And what's the harm in our enemies living in awe (and possibly fear of us), thinking we possess ESP or, at the very least, that we have eyes in the back of our head?

It is no coincidence that Para is dominated by the element Sky (sometimes written as *ether* or even *sound*) for hearing is the key to mastering Para.

Body-wise, Para/Sky controls the mouth, throat, and ears—our organs of communication. That's why the verb influence is "I speak."

Have you ever heard it said that, in order to solve a problem, you must "first give the beast a name"? What about, "Never speak the devil's name, lest you want him to appear!"

It's said one invites doom if an actor says the name of Shakespeare's *Macbeth* in a theater.

Others have even warned us, "A god defined is a god confined!"

All these sayings testify to the power of the spoken word. Just superstition? Perhaps. But it doesn't matter if you believe in a superstition, so long as your enemy does, and so long as you can use that superstition against him.

In previous books on Black Science we've set aside more than a chapter or two on how to use the spoken word against your enemies, while, at the same time, learning to guard ourselves and our loved ones from becoming victims to what is known as "word slavery." (See Lung & Prowant, 2001; Lung & Prowant, 2002, Lung, 2006.)

Learning to recognize and control the words that control us is one of the most important courses taught at the Black Science Institute. If you learn nothing else from your study of the Black Science, learn to really listen and thus to hear what other people are saying.

Mastering the art of hearing will allow you to seduce the fifth, and most important, of the Five Virgins.

NINE HOUNDS, EIGHTEEN LINKS IN THE CHAIN

"He cannot jump off his own shadow."
—Kannada Upanishad

Between the tenth and twelfth century of the common era, a revivalist movement known as Virasaiva swept across India.

Virasaiva means "heroic saivism" and refers to having faith in the god Shiva, the Destroyer.*

This movement resulted in social upheaval by and for the "have nots" against the "haves"; the disenfranchised poor and those trapped at the bottom of India's strict caste system, challenging the rich, privileged classes. According to one source "[I]t was a rising of the unlettered against the literate pundit, flesh and blood against stone" (Ramanujan, 1973:121).

In many ways this uprising was comparable to Europe's Reformation for, at its roots, it was fueled by religious revivalism.

Virasaivas wanted a return to what they felt was the original inspiration of their ancient tradition: direct communication with, and unadulterated worship of, Lord Shiva. Growing out of a fusion of conquering Aryan and indigenous Dravidian culture, Virasaisva followers were commonly known as "Lingayatas" since the phallus (Skt. *linga*) was the universally recognized symbol for Shiva. Lingayatas practiced a form of Bhakti (devotional) yoga. In this instance, devotion to Lord Shiva. Lingayatas also mixed in tantra and other forms of yoga with their Bhakti devotions.

The Vacanas saw the world as divided between Bhakta (devotees) and Bhavi (worldlings), men of faith and infidels.

Among the most devoted of devotees were itinerant Lingayat religious teachers/religious poets known as Jangama (moving men).

Lingayatas were sometimes also referred to as "Vacanas" after a type of religious lyric-poem style of the same name. The collection of these Vacana lyric-poems, sacred to the Lingayatas, is called the Kannada Upanishad (scripture).

Ligayata Vacanas desired to alleviate what they saw as stuck-in-the-mud Hinduism to a higher level, converting it from the practice of animal sacrifice to the sacrifice of the bestial self (Ramanujan, 1973:147).

They taught the common people through their poems (a convenient device since many peasants were illiterate). These poems were specially crafted, incorporating their mystic knowledge of mantras and verb influences specially designed to seize hold of a listener's attention, hold that attention—

---

*Hinduism 101: Brahma—the one God—has three "faces" he shows the world, three different aspects of himself: Brahma the Creator, Vishnu the Preserver, and Shiva the Destroyer.

literally spellbound!—and shepherd that person towards purification and enlightenment. (See page 31 and figure 6.)

Some Janagama became recognized masters (Skt. Prabhu). Especially gifted adepts were sometimes called "Siddhas," synonymous with the name used to describe the mystic powers they were thought to possess.

These *Siddha Prabhu* (literally, "power masters") taught a simple yet effective message of mystical union with Lord Shiva (aka "Ramanatha," and "Lord of the Meeting Rivers").

This mystic union could be accomplished through asceticism, meditations, and, for some, simply by listening to the specially crafted recitations, mantras, and Vacana poems.

Whether or not we buy into the idea that these Lingayat poets possessed "siddhas," special mystical powers, it is impossible to deny the influence they wielded on Indian history, on their peasant followers and kings alike.

At the very least this points to a kind of "charisma" or mystical "glamour" worthy of our investigation—be that power inherent and unintentional on their part, or deliberately sought after for personal Power.

The pounding of the swordsmith's hammer matters little the instant his steel passes close to a warrior's pounding heart.

Some of these power masters are known to history, although ten times that many are not—their brotherhood, *The Order of Red Robes,* unassuming in the least, mysterious and secretive in extremis. But at least they have left us their equally mysterious poetry.

In these poems we find not only "spiritual" insight, but also very practical, invaluable clues to the vanities and vulnerabilities of human nature.

One unique twelfth-century Siddha Pabhu was a woman named Mahadaviyakka. She didn't have much good to say about the inherent nature of humankind:

> Why do I need this dummy of a dying world? Illusion's chamberpot, hasty passions' whorehouse, this crackpot and leaky basement?

Cynical or just savvy? You make the call. Mahadaviyakka didn't hold out much hope for life and love in this world, she was saving herself for divine union with Lord Shiva.

Still she gave us more than one insight into human weakness. For example, her most famous poem:

Lust's body; site of rage; ambush of greed; house of passion;
fence of pride; mask of envy.

Yeah, bet she was a lot of fun at parties! Be that as it may, did you notice how closely her catalogue of human failings stacks up against the Five Warning F.L.A.G.S.?

Rage	Anger
Greed	Greed
Passion	Lust
Pride	Sympathy
Envy	Fear

Another Siddha Padhu known as Allama left us a similarly dim view of humankind:

Where the heart went I saw the brain run.
They don't know the day is the dark's face, and the dark
the days's.
The world tires itself thinking it has buried all shadow.

But most important, Allama alluded to the Nine Gates:

Closed and shut the bolts, of the nine gates and locked them
up, killed nine thousand men till he was left alone.

As every Black Science student knows, the Nine Gates is an oft-used metaphor in the East, one that symbolizes the nine openings in the body through which temptation can enter, by which we might attack our enemy: eyes, ears, mouth, nostrils, urethra (sex organ), and anus. (See Lung & Prowant, 2001, 2002.)

Twelfth-century Siddha Prabhu Basavanna (1106–1167) used similar symbolism when he warned us to beware the "nine hounds" of the uncontrolled mind and body:

Nine hounds unleashed on a hare, the body lusts . . . will my
heart reach you, O [Shiva] Lord of the meeting rivers, before
the sensual bitches touch and overtake?

Basavanna is perhaps the most intriguing of the Siddha Padhu. Unlike most of his more secretive brethren, Basavanna often took center stage.

A political activist and reformer, for a "holy man" Basavanna was often "condemned as zealot and conspirator by his enemies of whom he had many."

We sense a wee bit of Machiavelli—and maybe just a little Rasputin—in Basavanna. As his fame as Master grew, he became court adviser and eventually brother-in-law to Bijjala, king of the Mangalavada area of India.

Basavanna made enemies the way children make mud pies. He disregarded time-honored Hindu customs of sex and caste restrictions. More dangerous yet, he also rejected the complex—and often lucrative—religious rituals of his day.

His enemies counterattacked with gossip and accusations that first scandalized and then threatened to undermine the kingdom.

Finally Basavanna was forced to leave Mangalavada in order to prevent a civil war between those who adhered to his teachings and those whose liturgical livelihoods he threatened.*

Despite his hurried and harried exit stage left, Siddha Padhu Basavanna left behind him his many Vacana poems, chock full of intriguing observations on the frailty—and fatality!—of human nature:

What shall I call such fools who do not know themselves?

From Socrates to Sun Tzu, "know thyself" is job one. For the Black Science student, job two is to know your enemy *better* than he knows himself!

Does it matter how long a rock soaks in water? Will it ever grow soft?

In other words, tigers don't change their stripes.

When a whore with a child takes on a customer for money, neither child nor lecher will get enough.

A general can't fight effectively on two fronts.

The crookedness of the serpent is straight enough for the snake-hole.

---

*Virasaiva extremists eventually stabbed and killed King Bijjala, in part for sending Basavanna into exile.

Pick the right tool for the job—be that tool a blade, a bludgeon, a bully, or a buffoon.

> A grindstone hung at the foot, a deadwood log at the neck.
> The one will not let me float and the other will not let me sink.

Life screws us all in the end. Our job is to get to the Vaseline before the Grim Reaper does!

The bowstring is also a god.

"The bowstring is also a god" because it possesses the power of slavery or freedom, life and death.

A string is just a string and—like most untrained and unfocused people—useless . . . until you string it on a bow and fletch an arrow to it.

Thus, all things, no matter how seemingly innocuous, no matter how seemingly useless—people most of all!—all have a use . . . and *many* misuses.

One person who wasn't useless was the Lingayat monk who got the whole Vacana ball rolling back in the tenth century: Deavara Dasimayya, also called Salve Ramanath.

Deavara was a severe ascetic, into doing that whole fast-till-you-drop and then sleep-on-a-bed-o-nails thing. That is, until the Lord Shiva himself beamed down to tell Deavara to put away the pin cushion, scarf down a couple burgers (veggie, of course) and, from that day forth, worship only the sacred phallus.

Smart enough to know that when a god called the "Destroyer" gives you a suggestion, it might be a good idea to step lively, in short order, Deavara was a living testament to Shiva.

Deavara was also quite the martial artist, skills undoubtedly learned before he gave up everything to become a wandering monk.

This is not unheard of. Back in the sixth century, the twenty-eighth Buddhist Master, *Bodhidharma* (Ch. Tamo, Jp. Daruma) relocated from India to China where he helped found the (in)famous *Shaolin* monastery. There he taught the Chinese monks a form of exercise that eventually became known as kung-fu. (More on the origin of kung-fu and the mind and body training of Shaolin in the section on the red spears, in chapter 4.)

Likewise, there are many stories of Deavara's martial arts prowess.

He is said to have developed a martial arts discipline that allowed his body to absorb poisons his enemies used on several occasions to try to kill him.

Could this be the origin of the Chinese *Dim Mak* "Death Touch"? (See Omar 1989; Lung 1997; Lung & Tucker 2007.)

The tale is also told how Deavara once wove an enormously long turban cloth and that when a thief tried to steal it, a sharp wheel whirled out of it and slashed the thief.

This sounds suspiciously like a "flexible weapon" (e.g., nunchakus, chains, rope-weapons) similar to some later used by the Moshuh Nanren of China and the Ninja of Japan (Lung, 1997; Lung & Tucker, 2007).

We bring up Deavara's penchant for flexible weapons, because in his most well-known poem he uses the metaphor of "eighteen links" of chain, a chain used for binding, to represent the eighteen weaknesses all humans are susceptible to. The poem:

> You have forged this chain of eighteen links and chained us humans: You have ruined us O Ramanatha and made us dogs forever on a leash.

Notice how these all-encompassing "eighteen links" incorporate not only Mahadeviyakka's five failings, but also the Five Warning F.L.A.G.S., as well as all the Nine Gates.

During the course of any day, or lifetime, our mind bounces back and forth from one of these links to another, to another, like billiard balls in an earthquake.

But in the East they take it a step further, believing that these Eighteen Links are not just everyday wants and worries, rather these Eighteen Links are major traits, predispositions if you will, one of which predominates in each of us because of karma (past and present) and dharma (filial and social duty).

However, at any given time, our mind can embrace—or be overcome by—one of these Eighteen Links. . . . For example, you're driving along, racking your brain on problems from the past, present, or future, when you're suddenly cut off by some lead-footed jerk who adds insult to injury by flipping you a one-finger salute. In an instant your mind shifts gears, first to anger (14) and then to pride (17), as you begin to plot revenge!

But, as with all things in life, just because we have an "inclination" or

The Eighteen Links

1. *The Past.* Debts unpaid, trespasses unavenged: dark secrets kept: Blackmail worries.
2. *The Present.* Past & Future worries come together here; also, immediate problems: crisis.
3. *The Future.* Anxiety; feelings of inadequacy to pay future bills (see 6).
4. *The Body.* "The City of Nine Gates": potential illnesses and allergies. Health concerns.
5. *The Mind.* Mental illness; neurosis & psychosis; skewed mental filters (religion, etc.).
6. *Wealth.* Money problems.; desire for prosperity; get-rich-quick schemes (see 10, 15).
7. *Substance.* Position in life: peer pressure; social status; keeping up with the Joneses.
8. *Life.* Ageing concerns; midlife crisis: empty-nest syndrome.
9. *Self-Regard.* Paranoia. Looking out for 4,1; egoism in extremis.
10. *Gold.* Looks for opportunities; sucker for too-good-to-be-true offers; get-rich-quick schemes.
11. *Land.* Security-minded; concerns with home life and social circle: patriotic.
12. *Woman* (or man). Companionship; looking for love in all the wrong places; social.
13. *Lust* (see 4). Tendency to feel "passionate" about projects: leaps before he looks.
14. *Anger* (see 5). Susceptible to high blood pressure, stroke; road rage.
15. *Greed* (see 6). His greed may be rooted in insecurity & safety issues.
16. *Infatuation.* He sees it, he wants it; easily attracted to people and things.
17. *Pride.* Need for revenge; recompense: justice; (Asian) needs to save "face."
18. *Envy.* Jealousy; sees himself as victim of injustice: commiserate with his cause.

perhaps even a predisposition toward a particular vice or weakness, doesn't necessarily mean we have to act on that failing or frailty.

So whether our motivation is to guard our own mental borders or invade into our enemy's mental territory, we benefit from a deeper examination of each of the Eighteen Links:

The Past (1) makes us dwell on debts we have yet to pay (literal as well as figurative debts), and "trespasses" against us we have yet to avenge (study: Poe's *Cask of the Amontillado*).

Those "trapped" in the past are often staunch conservatives, traditionalists, and religious fundamentalists. They dream of a time gone by, one where they imagine themselves having a bigger, more potent part to play than that afforded by their present life.

We can deliberately bring the past back to haunt our enemies by digging up all their "dark secrets" they try to keep buried. These include birth secrets, body flaws, failure secrets, sex secrets, crime secrets, illness secrets, and death secrets. (See "Blackmail/Dirty Laundry List" in Lung & Prowant, 2006.)

The Present (2) has plenty of worries to keep us busy, from bothersome duties and obligations, to immediate unforeseen crises that demand our attention, drain our already rationed energies, and throw us off our all-important schedule.

To make matters worse, we can't seem to prevent our indiscreet past from rearing its ugly head, nor that terrifying hydra of the future from growing yet another hungry head.

When it comes to our enemies, our job is to make his matters worse!

The Future (3) never arrives, but that doesn't keep us from projecting yesterday's mess-ups and today's anxieties onto tomorrow and the next day, and the next . . .

Who doesn't have doubts or feelings of inadequacy when our mind races headlong into the future? Will I be able to pay my bills on time? Will I get that promotion? Will she say, "Yes"?

No matter how bright his future looks, our job is to show our enemy just how dim and dismal his future can become.

The Body (4) is susceptible to illnesses and allergies and all manner of discomforts, all of which we should be more than willing to help our enemy acquire.

An enemy, who is overly body conscious, i.e., dominated by the body, is likely to be a born hypochondriac—a neurosis we should go out of our way to reinforce in him.

And let's not forget our body is the "city of Nine Gates"; nine potential openings in our figurative—and literal—body armor through which an enemy can penetrate into our mind and set up shop.

Guard your own city of nine gates well, while diligently laying siege to your enemy's.

The Mind (5) is all we've got to work with—unfortunately. Mental illness, neurosis and psychosis, and more phobias than you can shake a stick at (unless sticks are one of your phobias?) plague us at every turn.

The brain is a delicate instrument, a wee bit of the wrong chemical this way or that—alcohol for some, LSD for others—and your enemy is soon babbling incoherently, channeling the spirit of his dead cat.

The human mind can also be confused and contorted by childhood drama and trauma, by cultic belief, and by other mental "filters"—some of which our enemy's parents and pastors were kind enough to provide him with, the rest of which we'll be glad to help him acquire.

Wealth (6),money problems, tax us all. Even Bill Gates has money problems. . . . Of course, his money problem is that he makes it faster than he can give it away. I know what you're thinking: We should all have such "problems"!

Beyond simple day-to-day money problems, of making a living till we finally die, many people put in extra-long hours trying to get rich. Still others are obsessed with getting rich.

Let's hope your enemy is obsessed with wealth since this is the perfect opening for you to hit him with your latest "no money down!", "interest free!", "get rich quick!" scheme.

Substance (7) refers to your position in life, where you see yourself standing in the "great pecking order." A true "man of substance" never frets over such things since he is confident of his place in the grand skein of life.

Other, not so confident people, obsess over their social status. This sort of person is susceptible to peer pressure (yeah, just like in high school!). They have to "keep up with the Joneses," they have to have a new car, a new boat, simply because the neighbor or the guy in the next cubicle over has one.

If this sounds like your enemy, show him how you can help him get one up on his neighbor or be the envy of all his coworkers and he'll be eating out of your hand.

Or, seed his mind with thoughts of how's he's *not* keeping up with the Joneses, tempting him to get in over his head in debt by taking a fool's chance in Hell.

Life (8) kicks all our asses in the **end**, right? Aging concerns haunt many, and not just the elderly and the vain.

Eric Erikson mapped out several stages of human development, each stage possessing a psychological brass ring we need to grab hold of if we are to maintain good mental health and a feeling of self-worth (see Lung & Prowant, 2006).

Our job, of course, is not only to prevent our enemy from grasping that brass ring but, ideally, to make him over-reach himself so far he falls all the way off the merry-go-round of life!

Age concerns happen as early as puberty, when teenagers are striving to establish their unique identity. At this age they want to look older, and act older (by drinking, smoking, having sex, etc.), all the stuff adults do. That's why this is the perfect time for gangs, cults, and online predators to snatch kids up. At this age, kids will listen to anyone who pret**end**s to listen to them.

As we grow into our twenties we are expected to move out on our own, get a job or an education, start a family. These are expected "milestones" on the road to "maturity" and "independence." Of course, we often feel like failures if we fail to live up to these societal expectations.

According to Erikson, this carries on into middle age—when women experience "empty nest syndrome" (a feeling of uselessness after all her children have grown up and moved out). And middle-age men often hit what's known as a "midlife crisis," where fast new cars and fast young girls catch their attention.

Elders suffer through age difficulties as well, not the least of which is what Erikson calls the "integrity vs. despair" stage, when a senior begins to look back on his life to take an accounting of what he's accomplished. Confidence men know that this vulnerable period is the perfect time to target wealthy seniors. Our job is to keep our enemy from getting a grip on that brass ring.

Self-Regard (9). There's nothing wrong with looking out for number one—yourself. Except when we lose sight of the fact we share this world with another six billion human beings—some of whom just might prove useful to us!

Gold (10) in this instance refers to someone always on the lookout for that "golden opportunity," a chance to move ahead. This isn't necessarily a bad thing in and of itself, only when it becomes an obsession and/or when there is a total disregard for the welfare of others.

While often ruthless, this type of personality is just as likely to be a sucker for "too good to be true" offers and "get rich quick!" schemes. In this way he is similar to those dominated by *Self-Regard* (9). However, whereas someone dominated by self-regard might be smart enough to see how allying himself with others of like mind might be of benefit to him, the *gold*-dominated personality is blinded by his "gold fever" and refuses to even consider an alliance of convenience, too paranoid and myopic to see how such an alliance might benefit him in the end.

This kind of person is the easiest to turn to your purposes since his only loyalties are to himself and to money—in his mind, one and the same.

This kind often make good spies, although, when employing such a spy, take special care to insure you are always the one holding the purse strings or else he will leave you holding the bag by selling you out to the highest bidder.

Land (11) equals security. There was a time—before the United States Supreme Court decided *eminent domain* meant you really don't own your own home anymore, when owning your own home was truly the American Dream. On the purely physical level so many people function at, owning land (and material goods and property of all sorts) allows them to see themselves—and be seen by others—as being "prosperous" and "wealthy," of having economic and social stability.

Beware the old adage: The more things you have, the more things have you. And the more the chance a wily enemy will find a way to use those possessions against you.

Psychologically, *Land* includes such things as family, clan, and even country. The die-hard patriotic fall into this category (see 15, below).

Woman (12) (or *Man*, depending on your gender, inclination, and how many beers you've had) concerns companionship, from friends and familiars, to spouses and "significant others." The desire for companionship is strong in all of us, from social get-togethers to intimate pairings.

Those desperate (or just horny) enough are often caught "looking for love in all the wrong places."

Remember the rule about never going grocery shopping while hungry? The same rule applies to looking for love.

Let's hope your enemy is one of those who does look for love in all the wrong places. First you can encourage him to look for love in all the wrong places—perhaps even setting him up with it . . . before then sending the pictures to the *National Enquirer*.

Lust (13) is not just physical lust. This predisposition encompasses the (often uncontrollable) tendency to feel "passionate" about some thing or some person. That "thing" can be a religious or political cause, hence "rebels" and terrorists are usually dominated by this *Lust* category.

This personality type leaps before he looks (see 4).

Anger (14)–dominated people are susceptible to high blood pressure, stroke, and fits of "road rage."

Sun Tzu talks often and at length on what a godsend it is to have an enemy who is easily provoked to anger. Making an enemy see red is the surest of ways of getting his red on your blade! (See 5, 17.)

Greed (15) as used here is different from the obsession with (physical) gold and (psychological) wealth. This greed is rooted in insecurity and safety issues. This personality type can never get enough. He surrounds and insulates himself with "walls" constructed of physical possessions and psychological coping mechanisms (e.g., rationalization). To him, these insulating layers represent security from want and worry, and sometimes from responsibility.

The best example of this type of person is Adolf Hitler. The future dictator lived many years of his early life with his mother during a time when their home life was precarious and their future uncertain. Later he would spend a dismal period as a starving artist in Vienna before the outbreak of World War I. After the war, he continued to agonize over his future as he was bandied about a post-war Germany on the brink of civil war.

But, finally, as he began to accumulate political position and power, he began to figuratively, and in some cases literally, build walls to insulate himself against poverty and powerlessness.

Thus, if we look at Nazi Germany at the height of Hitler's power, we find a succession of concentric "security" circles, leading ever inward to where Hitler himself sat, insulated from abuse, poverty, and hunger as he never was during his youth.

To approach an enemy like this, one who builds "greedy" walls around him, yet who can never seem to get enough, we either offer to help him build those walls—thicker and higher until they become his prison!—or else begin slowly chipping away at his safety and security insulation.

As with Hitler, so with our own enemy: As his insulating security walls begins to crumble one by one, and the outside world of reality draws ever closer to the center where he sits, so too his confidence and ultimately his sanity will begin to crumble (see 6 and 10).

Infatuation (16) means he sees it, he wants it. So if it's something you want him to have in his life, show it to him. Just make sure it has plenty of flashing lights attached since this type of personality is visually oriented.*

Being visually oriented, he is easily distracted. Short attention span, he falls in love quickly (with people, places, and things) and just as quickly loses interest.

We are forever trying to distract our enemy, luring his mind away from important things he should be paying attention to, on to the more supererogatory.

Thus our strategy is attract him to distract him.

Pride (17) goeth before a fall, it's often been said. This kind of pride is not the kind of pride one feels when their son makes the winning touchdown, or their combat unit wins a decoration. This type of pride is the *ego-pride* that takes offense too easily and too often. The pride that longs for revenge and recompense.

More often than not, these people convince themselves that the vengeance they're seeking is rooted in "justice," but that is often just the excuse used to put the bullets in the gun.

In the Far East, this is called saving face, or regaining face after a failure or an insult. To Asians, "face" is synonymous with "honor."

In America, the street vernacular for redressing one's wounded pride is "getting your manhood back."

Convince him he's been "punk'd out," made to look the fool, and he'll already be headed out the door, rage in one hand, revenge in the other (see 14).

---

*The three types of orientation: Visual ("watchers"), auditory ("listeners"), and tactile ("touchers") are described fully, along with strategies for dealing with them, in *Black Science* (Lung & Prowant, 2001), and *Mind Manipulation* (Lung & Prowant, 2002.)

Envy (18) is the twin briers of hate and greed firmly rooted in jealousy. This type of personality is what's called a hater. He hates everybody and everything—God, fate, and the universe included—because others get ahead while his genius goes unrecognized.

He sees himself as a victim of injustice; passed over for promotion and unappreciated. Sun Tzu points to this personality type as a prime candidate for recruitment as a spy.

Commiserate with his cause. The world doesn't recognize his genius, his uniqueness, so as soon as you do, you'll become his bestest new friend. Soon he will become hungry, even dependent on your praise. You can then direct his envy (hate) against another of your enemies. Two birds, one stone.

In the West, the number "13" is considered unlucky. In many places in the East, it's the number "4."

But, I can guarantee that unless you take the time to learn these various personality types, these inherent weaknesses we all share, "18" (as in *the Eighteen Links*) might prove the unluckiest number of all!

THE DARK SIDE OF DHARMA

So much of the so-called New Age mysticism and philosophy comes from ancient Asian schools of thought. This includes yoga, Zen and other forms of meditation, Tantric sex practices, as well as Tibetan esotericism and wisdom straight from the smiling lips of the Dalai Lama.

In the past, intrepid Western seekers like Helena Blavatsky literally braved death to penetrate into the wilds of Asia. Her Theosophy school is deeply indebted to Asian mysticism for most of its philosophy. Likewise, American Master Alan Watts devoted much of his life to helping the West unravel the riddle of Zen—and almost succeeded.

And, of course, the Beatles turned us all onto TM—transcendental meditation.

Sometimes there is a direct line of transmission from Eastern Master to eager Western neophytes. The Japanese call this "ketchimyku"—the blood pulse.

Other times these teachings are "filtered" through an interpreter—perhaps a lifelong devotee and trusted disciple. Other times such "sacred wisdom" turns out to be only snippets of sayings or half-learned lessons stolen and then parroted by some two-bit fakir.

Critics are quick to point out how this allows for too much to be "lost in the translation" from East to West; that this opens up the potential for unintentional misinterpretation and even intentional misuse even from the most harmless of teachings.*

However, as with most things in life, and all things Black Science, there is always the potential of perversion—that the simplest, most innocuous and harmless of words and rituals, can be used for evil.

There is always the potential for harm. That's why we study. So that that harm won't be done to us and our loved ones.

In India, the word "dharma" is best translated to mean "duty"—your duty to yourself, duty to your friends and family, your place in, and duty to society as a whole.†

Dharma can also mean your "destiny," understood as fulfilling one's duty within India's caste system, and to the karma that has brought you to this particular point and passage in life. (Read: The *Bhagavad-Gita.*)

As a Black Science student, think of it as your "dharma," your duty to study any and all ideas and insidious plottings, higher concepts and low-life con games, any or all of which might potentially affect—for well or ill—yourself and your loved ones.

Who knows, this might even be your dharma—destiny!

BUDDHIST BLACK SCIENCE

"All that we are is the result of what we have thought, it is founded on our thoughts, it is made up of our thoughts."
—The Buddha

No other Eastern school of thought has had as much influence on the Western New Age movement as Buddhism.

Guatama Siddhartha (560–477 BCE) was a prince of northern India who gave up his throne to go in search of enlightenment, which he found

---

*Hinduism, and her stepchild Buddhism and Jainism, all adhere to the principle of "Ahimsa"—do no harm, the principle most promoted by Gandhi.

†*Dharma* written with a small-case "d" is used to refer to the Hindu concept of "*dharma*"; whereas Dharma with upper-case "D" is used as a synonym for the teachings of the Buddha. *The "Dharma"* always refers to the teachings of Buddhism.

after forty years of study and struggle, earning him the title "Buddha" ("one who is awake").

Today the Tibetan Buddhist leader the Dalai Lama has become a world-recognized spokesman, not just for Tibet, but for Buddhism worldwide.

Likewise Buddhist celebrities like singer-actress Tina Turner, actor Richard Gere, musician Herbie Hancock, and martial artist–turned movie star Steven Seagal have all helped promote Buddhism in the West.

And who doesn't remember, "Snatch the pebble from my hand, Grasshopper," from the '70s *Kung-Fu* TV series?

Kung-Fu spotlighted the Shaolin Order of monks, a fusion of indigenous Taoism and the Buddhism brought to China from India around 520 CE. How many Westerners were influenced to check out Buddhism by this show? And by the Bruce Lee "kick-flick" craze around the same time. In fact, many Westerners were introduced to Buddhism through their martial arts study.

But not everyone is pleased by the welcome reception Buddhism has received in the West. Some try to hint at a darker side to Buddhist teachings, finding connections between Buddhism and shadowy Freemasonry groups.

It is true that through the study of the teachings of Buddha insights into the nature of human beings are inevitably revealed, revelations that, in the wrong hands, could all too easily be used to manipulate others.

The same blade that cuts the dinner roast can make a meal of the heart as well.

Efforts to prevent the evil perversion of Buddha's teachings are at the heart of the myth(?) of *The Nine Unknown Men*. The best version of this story was popularized by French diplomat and travel writer Louis Jacolliot (1837–1890). While traveling in the East, Jacolliot uncovered the tale of how a secret group of Buddhist scholars, enlightened men all, were chosen by Asoka, the third-century Buddhist emperor of India. These nine men, whose identities must remain secret for obvious reasons, took a blood-vow to use their wisdom (and, by default and necessity, backdoor machinations), to secretly rule the world—albeit benevolently (Picknett & Prince, 1999:264).

When one of these Nine Unknown Men passed on, another would take his place from a pool of worthy candidates, so there would always be Nine Unknown Men somewhere in the world.

These original nine men traded in their traditional orange Buddhist robes for black robes, signifying their dedication to remain "unknown" behind the "Black Curtain" and to serve out their life in secret service to mankind.

At first glance there's a seeming contradiction between Buddhism's pledge of *"Ahimsa"* (harmlessness) and Eastern warrior cadre like samurai and shinobi ninja warriors calling themselves Buddhist while hacking their way down through history. The good news is Buddhists don't start wars (though Buddhist samurai and ninja have been known to finish a few!).

However, unlike some religions, where the "faithful" have "religious police" watching over their shoulder 24-7 to insure they stay the "faithful," in Buddhism it *is* left to the individual to decide how best to follow Buddha's "middle path."

The Buddha left behind only an outline for finding freedom and enlightenment in this world, his Four Aryan Truths:

> *The nature of life is suffering.*
> *Suffering is caused by desire.*
> *Desire can be overcome.*
> *Desire is overcome by following the Eight-Fold Path.*

Buddha's Eight-Fold Path consists of having Right Views; Right Desires; Right Speech; Right Conduct; Right Livelihood; Right Endeavors; Right Mindfulness; Right Meditation.

What makes Buddhism unique is that one Buddhist can't tell another Buddhist how to apply these eight principles. As a result, a thousand sects and schools of Buddhist thought have emerged down through the centuries, each with their own particular interpretation of, and techniques to accomplish, Buddha's ideal.

Since, at its most basic Buddhism delves into the depths of the human mind to uncover the roots of a person's desires—our toxic attachments to the world—it is not surprising such intimate knowledge of the human mind and character holds the potential for misuse.

For example, what might the Black Science student glean from the *Dhammapada*, the 423 sayings attributed to the Buddha himself, sayings that form the essence of Buddhism:

Weakness and Failings

"Life is easy for a man with no shame, for a man who is bold like a thieving crow, and for a troublemaker who is insulting, arrogant, and wasteful." (Saying 244)

"As rain seeps through a badly-patched roof, so too lust leaks into the untrained mind." (13)

"Senseless and foolish men sigh into sloth, while the wise man guards sincerity as his greatest treasure." (26)

"Seek not after vanity, nor the enjoyment of love, nor lust. Only a man who meditates on sincerity obtains final joy." (27)

"There is no fire like lust, no spark like hatred, no snare like folly, and no torrent like greed." (251)

"The foolish man, obsessed by riches to the exclusion of all else, soon destroys himself, just as surely as if he were his own worst enemy." (355)

" 'He abused me, he beat me, he robbed me and so defeated me,' hold tight to such thought and your hatred will never cease." (4) "Fields of grain are damaged by weeds. So too man is damaged by lust (356) . . . by hatred (367) . . . by delusion (358) . . . and by craving (359)."

Black Science: If we can be "damaged" by these four weaknesses, so can our enemy.

Know Yourself

"The fault of others is easy to see, but our own faults are often difficult to admit. This is why a man can so easily separate his own faults from his neighbor's faults, like separating wheat from chaff. All the while he hides his own faults the way a dishonest dicer hides his snake-eyes." (252)

"Even if a man is quick to spot the faults of others, and quicker still to take offence, his own evil faults will continue to grow until they lead him to destruction." (235)

Black Science: We spot our enemy's faults, we help them grow, we shepherd him to a speedy demise.

"Realizing that this body is as fragile as glass, first make your mind firm— like a fortress—and then cut through the falsehood of this world with the sword of knowledge. Watch your enemy even after he is brought down, and never falter. (40)

"Whatever one hater can do to another hater, whatever one enemy can do to another enemy, still your untrained mind can do you even greater harm." (42)

Conquer Yourself

"The warrior who conquers a thousand times a thousand men in battle is not as great as the man who conquers himself. That man is the greatest of warriors." (103)

Black Science: He who overcomes others is strong. He who overcomes himself is mighty. Echoes of Sun Tzu. To know yourself, your enemy, and the battlefield are the three musts for any strategy. But the greatest of these *is* to know yourself.

"Become an island. Exert yourself. And promptly be wise!" (236)

"Ditch-diggers direct water wherever they like. Archers bend the bow to their will. Carpenters stretch a log of wood to their liking. In the same manner, the wise create themselves." (80)

Black Science: There comes a time when the dramas and traumas of our child-hood, the crisis of our youth, all our past indiscretions, our present fears and future anxieties, all must bend their knee to our will and determination to remake ourselves—free from the fetters of the past, confident against the challenges of the day, forging the future to suit our new self.

"When there is something to be done, do it! Attack it vigorously." (313)

Black Science: "Attack it vigorously" . . . but always with concentration and attention to detail. Because:

"Even a grass-blade, 'if badly grasped, cuts the hand . . .' " (311)

Black Science: Having conquered yourself, can conquest over your enemies be far behind?

Control Your Senses

"Even the gods envy a man who controls his senses as adroitly as a teamster controls his horses. This man's senses have been subdued. He is free from pride and from evil intent." (94)

"Cut off the Five. Renounce the Five and cultivate five more. A monk who escapes from the Five Fetters is praised as 'One who has crossed over the flood.' " (370) (See *Jing Gong* in chapter 4.)

Black Science: Control your five senses and then use your enemy's "Five Fetters" to control him!

Control Your Anger

"Control the anger of your body. Guard yourself against the temptations of the body and practice virtue with your body." 231)

"Control the anger of your tongue. Guard yourself against the temptations of your tongue and practice virtue with your tongue." (232)

"Control the anger of your mind. Guard yourself against the temptations of your mind and practice virtue with your mind." (233)

"Remain steadfast, as one who has learned to control body, tongue, and mind. Such a man *is* indeed well-controlled!" (234)

Black Science: Your anger controlled is your enemy's worse nightmare. Your enemy's anger out of control, is the answer to a dream.

Practice Restraint and Patience

"Restrain your eye. Restrain your ear. Restrain your nose. Restrain your tongue." (360)

"Restrain your body. Restrain your speech. Restrain your mind. Restraint in all these is the key." (361)

Black Science: First we restrain, then we retrain.

"Patiently I endure the slings and arrows of abuse from an ill-mannered world the way a noble battle-elephant endures annoying arrows." (320)

Black Science: We practice patience and restraint so as to better cause our enemy to lose patience and end up in restraints!

Guard Yourself

"A well-guarded fortress, defenses inside and out, that is how a man is to guard himself." (315)

"With sincerity as your watchman, guard your thoughts well. Raise yourself above evil the way the noble elephant pulls himself from the suck of the muck." (327)

"Hold yourself dear by guarding yourself dearly. The wise man keeps watch himself at least one of the three watches of the night." (157)

"Guard well your thoughts though they be difficult to perceive, crafty, and running about here and there. Thoughts well guarded bring joy." (36)

"This mind of mine formerly wondered and wandered wherever it wanted and did as it pleased. But from now on I will control it perfectly, the way a mahout controls his beast with a sharp hook." (326)

Of Friends and Fools

"Fools are their own worst enemy. With little understanding they plant evil deeds which all too soon bear them bitter fruits. (66)

"Drop-by-drop water fills the pot. In the same way, the fool becomes full of evil even though the fool, supping from it little-by-little, declares to himself, 'Evil cannot intoxicate me!' " (121)

"Do not have evildoers for friends. Do not have low men for friends. Make only the virtuous and the best men your friend." (78)

"If you meet a wise man, one who is quick to detect true faults and who places blame on what is blame-worthy, follow that fellow, for such men are the rarest of treasures." (76)

"While on your journey, if you do not meet with one who is your better, or at least your equal, keep your solitary counsel. There is no companionship with a fool." (61)

SACRED BULL OF THE BHAGWAN

"Confusion is my method. The moment I see you accumulating something, creating a philosophy or theology, I immediately jump on it and destroy it."
—**Bhagwan Shree Rajneesh**

Despite its enormous potential for helping us plumb the depths of the human psyche, helping us better understand ourselves and others, some still imagine an inherent sinisterness—something akin to anarchy!—behind Far

Eastern philosophy in general and the West's growing interest in all things Eastern in particular.

In fact there's some who never tire of finding connections—no matter how ephemeral—between shadowy figures East and West, Hitler being a favorite (Ostrander & Schroeder, 1970):

> Hitler talked of Ultima Thule, the magic center of a vanished civilization; of Shamballah, the legendary underground camp in the Himalayas whose forces of violence and power control humanity; of Agarthi, another legendary underground Himalayan city of goodness and meditation. (p. 327)

And (Pauwels & Bergier, 1971):

> Hitler was apparently trained in mediumship by Professor Hanshofer of the University of Munich, who is said to have been initiated, during a stay in Japan, into one of the most important secret Buddhist societies.

Still others point to how Asian techniques of Yoga and mind control received more than passing scrutiny during the heyday of psychic research in the Soviet Union. Take for example the experimental "Suggestopedia" school of mind expansion and mind control developed in Bulgaria by a certain Professor Lozanov:

> Lozanov's suggestopedia, taken from yoga, is a genuine form of mind expansion. It's revolutionary—evolutionary, perhaps. What powers, what talents will surge into being as more of the mind is freed? (p. 300)*

There have been indications that such "forbidden" study has not only continued long after the end of the Cold War and the fall of the Iron Curtain, but also has actually escalated; a new sort of mind control "arms race," now hidden behind the Black Curtain in Asian ports of call. (More on this in chapter 5.)

Shades of Dr. Fu Manchu! One can't help but suspect prejudice when

---

*The particular type of yoga utilized by Lozanov was "savasanna," a specialized yogic relaxation technique (Pauwels & Bergier).

these naysayers point out the mysterious power these "strange" and "inscrutable" Asian philosophies have to entice and entrap the unwary Western mind.

Such critics often point with alarm to the inroads Asian "cults" have made into America and the rest of the West. Cults like the South Korean–based "Moonies" and the ubiquitous Hare Krishna movement.

Nowhere has such criticism of Asian cultic methods of mind control and manipulation been so suspect (and later proven) than with the cult of the Bhagwan Shree Rajneesh.

The Cult Leader

Bhagwan Shree Rajneesh was born in India on December 11, 1931. He was born into the Jainist religion, an off-shoot of Hinduism devoted to Ahimsa nonviolence.

According to Rajneesh's official PR, he had his first transcendental religious experience—called *samadhi*—at age seven, and continued to have them all his life. Despite these religious "revelations," Rajneesh went on to receive his master's degree and continued working as a university professor while laying the foundations for his future "cult"—balancing academe with the arcane.

Finally, he resigned his university post in 1966 in order to become a full-time spiritual leader.

By 1974, he had formed the organization that would later become his "cult," humbly calling it the Rajneesh Foundation.

In 1981, Rajneesh moved his growing cult from India to New Jersey, and then farther west to the small town of Antelope, Oregon. Around this time the media started noticing Rajneesh's setup after complaints surfaced that Rajneesh followers had begun pouring into Antelope, Oregon, at an alarming rate, in short order overrunning the small town. Over 7,000 followers visited in 1982 alone.

By December 1982, Rajneesh followers had commandeered the local elections, winning the Antelope mayor's office and a majority of seats on the city council. They promptly renamed the city Rajneeshpuram.

Allegations soon surfaced that Rajneesh followers had used intimidation, voter fraud, and even a failed attempt to use botulism to poison the town's popular salad bar to prevent longtime Antelope residents from voting.

By 1984, Rajneesh claimed 250,000 followers worldwide, between 10,000 and 20,000 of those in the United States.

Eventually, a court of appeals nullified what they determined was a hijacked, jury-rigged election and restored Antelope back to its original citizens. A moot point, since, by then, the Rajneesh empire had already begun to fray around the edges.

In 1985, Rajneesh's main lieutenant was convicted of several felonies.

The media—and finally, authorities—began to question what an Indian holy man, one who had supposedly taken a vow of poverty, needed with a fleet of limousines. More ominous still, why was a religious group supposedly dedicated to nonviolence seen toting around automatic weapons?

Eventually, Rajneesh himself *was* deported and died not long after in his native India.*

The Cult Method

It has been argued that Westerners just don't "get" Rajneesh—or other Eastern cults for that matter. Case in point, Westerners don't understand Indian religious tradition governing the master/disciple bond.

According to Indian ecclesiastical protocol, to reach ultimate enlightenment a student is expected to surrender himself to a recognized master— who is seen as not only his superior in spiritual matters, but who is also the earthly embodiment of the Divine, acting in the god, or goddess' stead. Often this requires the seeker to divest themselves of worldly wealth, and to make a gift of their worldly possessions to their new master (who in theory then passes this wealth along to the truly needy). . . . At least this was the explanation given for Rajneesh's fleet of limousines.

Those automatic weapons his disciples were seen brandishing? Guess they're still working on that one.

Be that as it may, from a Black Science perspective, we come to bury Caesar not to praise him. It is the methods Rajneesh reportedly used to draw, stall, over-awe, and finally overhaul his followers that interest us.

---

*At Rajneesh's cremation, one of his wives committed suicide by throwing herself onto his pyre, an ancient (outlawed) Hindu custom known as "suttee."

Tailor Your Message.

Like any successful cult leader, or any good speaker for that matter, Rajneesh crafted his talk(s) to fit his immediate audience. In other words, he used tactics, techniques, and trigger words specifically designed with the needs (and vulnerabilities) of the individual in mind.

Switching spiritual horses in midstream may seem contradictory and confusing but it keeps your followers off balance, doesn't let them catch their breath, and keeps them from becoming too "cocky" since they are always having to guess how best to please the Master today. This also encourages competition (if only on a subconscious level) between disciples, all vying for the lion's share of the Master's attention—each willing to go to greater and greater length (up to and including murder?) to gain the Master's favor.

The Play's the Thing.

Any cult leader worth his salt has be a constant and consummate showman—in other words, he has to keep the followers "entertained," or at the very least, perpetually enthralled (Gordan, 1987):*

> If there were underlying principles and an operative technique at work in Rajneesh's counsel they were, in fact, contradiction, confusion, and continual change. . . . He would often tell one man to meditate more and the next that he meditated too much. People who craved structure were told to abandon it and those who fought against it were forced into it. . . . Contradiction, confusion, and continual change—Masters have always used these techniques with their disciples. They create receptivity, and openness to seeing things *as* they are, *as* they break habitual patterns. (p. 58)

These "3 C's" are the con man, the congressman, and cult leader's best friends.

You gotta think like a cult leader: The more contradiction and confusion in the parishioner's mind, the more coins in the collection for the preacher.

---

*"Enthralled," from the root "thrall," meaning "slave." Synonyms include servitude, bondage, and slavery.

FYI: On the subject of passing the collection plate, on how much the preacher is permitted to keep for himself and how much should go into the church coffer? Having placed all the tithes collected onto one plate, the preacher suddenly tosses all the money up into the air.

The parts that God wants to keep, he keeps. Anything that falls back to earth the preacher keeps!

Look into My Eyes . . .

In what has to be the definitive study on the Rajneesh cult, *The Golden Guru: The Strange Journey of Bhagwan Shree Rajneesh* (1987), author James Gordan points to the fact that ancient Tibetan Masters sometimes used this same "3 C's" method—contradiction, confusion, and continual change—as have modern Western hypnotherapists, to induce a hypnotic trance. No surprise Rajneesh was no stranger to hypnotic techniques:

> In discourses and darshans [one-on-one talks] Rajneesh also used hypnotic techniques to bypass his disciples' conscious defenses, to win their assent to his words, and to enhance their "transference" to him. He created confusion and elaborate paradoxes and contradictions, which baffled their rational minds and habitual ways they looked at themselves and the world. He used his voice, varying the volume and pace of speech, punctuating, modifying his words with his hands and eyes, even with his stillness. As his disciples listened and watched their minds slowed. They followed the winding, discursive thread of his stories the way the eye follows the motions of a tiny falling feather. In trance they were more receptive and suggestible. (p. 235)

Setting the Stage

A good cult leader, like any good Broadway director, first sets the stage. He adjusts the lighting, bathing some parts of his stage in brilliance, plunging other parts into abysmal dark, reflecting the chiaroscuro in his own mind. He manipulates the scenery and background against which his rigorously rehearsed actors will dance to and fro to his complex choreography, as they all the while recite flawlessly performance after performance line for line the roles he's written for them.

At his cult HQ in India, and later at his compound in Antelope, Oregon, Rajneesh masterfully manipulated the setting to maximize its effect on both the faithful and on future followers.

For all the talk of "openness" and "freedom," at the Rajneesh compound security barriers and checkpoints were everywhere. This included electronic scanners and posted guards, giving the impression one was entering the presence of a king or president.

Cult-think: The more important the Master is, the more important we, his followers, are. Important people are always in danger, ergo, all this security means the Master is very important.

FYI: Cults often use extremes of "security" to feed into their followers paranoia that they are being "watched" and "persecuted" by outsiders.

Elaborately staged entrances and exits were arranged for Rajneesh, leading up to his perch atop a raised platform, above the faithful, so that they literally had to look up to the Master.

The compound overall was mapped out and manipulated to present an idyllic setting, away from the troubles of the outside world (i.e., reality!). Inside this perfect world, the tenets and principles of the cult leader seemed to work. In this secluded "temperature controlled" environment, the disciple begins to believe in the message of peace (etc.) taught by his master.

Within the Rajneesh community, distinctions of age, class, nationality, race, and even specific religions no longer mattered and so soon became blurred in the mind of the disciple.

Your past deeds and indiscretions no longer followed you, any future anxieties you might have brought with you vanished, as the future ceased to exist.

Children were allowed to play without parental censure. Teenagers were treated like adults. Adults themselves were relieved of responsibilities and worries foisted on them by "the outside world." Here they could indulge their fantasies and explore their fears safely, the community itself acting as mother. Rajneesh playing the omnipotent father figure.

Out with the Old . . .

By breaking down all of a disciple's traditional reference points (age, race, nationality), Rajneesh in effect destroyed that disciple's identity. This in preparation for providing the disciple with a new cult identity.

. . . In with the New

After the "break-down," the systematic dismantling of the new disciple's old identity, now comes the "build-up." The same meditation techniques and therapy groups that "helped" new disciples "let go" of their old "attachments" (i.e., identities) also helped the disciple adopt their new cult identity.

Hugs all around. Much tears and blissful acceptance, safe in the loving arms of an idyllic community, secure under the omniscient gaze of the Master.

Constant dictates and directives direct from the mouth of the Master, and his always watchful, ever attentive lieutenants and instructors provided all the guidance the neophyte needed to embrace the cult's worldview.

Along with this new worldview comes new clothes (loose-fitting robes meant to further blur gender distinctions, body advantages or disadvantages, and previous social status).

FYI: Cults often dress their new disciples in robes (and other cult-distinctive clothes) for three very practical reasons: (1) like the military, everybody now looks alike, helping break down the individual identity, (2) wearing "funny" clothes further alienates the cult member from "normal" people around him, and (3) in the case of actual robes, it's a lot harder for a new disciple to run away in a "dress" than in blue jeans!

If you were really special you were given a prayer book, prayer beads, or holy icon-pictures touched by the Master himself. The disciple now has a new name, a new identity, in a perfect new cult-controlled setting where everything he sees and hears makes perfect sense.

If this "break 'em down" to "build 'em up" cult indoctrination sequence sounds familiar, then you've probably read the chapter on "Brainwashing" in *Theatre of Hell: Dr. Lung's Complete Guide to Torture* (Loompanics, 2003).

EAST INDIAN CULT
"The Rajneesh Foundation"
(aka "Rajneeshees")

Founded: Mid-1980s, India.

Parent Religion: Hinduism, Jainism, Misc. East Indian schools of philosophy.

Original Leader/Founder: Bhagwan Shree Rajneesh (b. 1931)

Current Leader: Misc. splinter groups, after Bhagwan Shree Rajneesh's death in late 80s.

Ancestor Claims: Ancient Aryan Hinduism, Jainism, eclectic philosophies.

Special Mandate: Establish worldwide peace under "divine" leadership of Bhagwan.

Cult Identity: "The Chosen Ones", "Enlightened", "Disciples", "Children".

Primary Target Recruits: Middle-class, White "yuppies," European and American.

Holy Book(s): Various Rajneesh writings, snippets of Hindu, etc.

Membership: 250,000 worldwide at height in mid-1980s. 20,000 USA.

Special Teachings: Bhagwan Shree Rajneesh is the "promised one," fully enlightened Master reincarnated on Earth to usher in a period of peace and prosperity under his benevolent & enlightened dictatorship.

(See Sacred Bull of the Bhagwan earlier in this chapter.)

Figure 7.

HOME-GROWN AMERICAN CULT
"The Nation of Islam"
(aka "Black Muslims"/"Farrakhaners")

Founded: Detroit, MI 1931

Parent Religion: Islam, Moorish Science Temple.

Original Leader/Founder: Elijah Poole (1887–1975) aka "Elijah Muhammad"

Current Leader: Louis J. Farrakhan (b. 1934)

Ancestor Claims: Outwardly Islamic, inner teachings tell of extraterrestial origins of the Black race.

Special Mandate: Establish separate Black "Nation" within the current borders of the United States, based on doctrine of African racial purity & superiority.

Cult Identity: "Original Asiatic Black man," "Fruits of Islam," splinter groups "Death Angels"

Primary Target Recruits: African Americans.

Holy Book(s): Koran, misc. writings of Elijah Muhammad and Farrakhan.

Membership: Estimated 80,000

Special Teachings: Caucasians are the result of an evil genetic experiment conducted by a "big head scientist" on the isle of Patmos.

There is a gigantic spacecraft known as "the Mother Wheel" (built by superior African technology) in orbit around the Earth and/or hidden on the dark side of the moon (where the "original Asiatic black man" came from 66 trillion years ago).

Figure 7 *continued.*

I'm sure you realize that the cult tactics used by Bhagwan Shree Rajneesh are in no way exclusive to cults of Indian extraction in particular, or even Far Eastern cults in general.

No, these methods of mind control and manipulation are universal and have been used down throughout history, in every clime and time, by unscrupulous cults and cadre with varying degrees of success—that particular degree of success directly proportionate to the ruthlessness and/or religious fanaticism of the particular cult leader(s).

In fact, placed side by side, it's easy to see how cult lures, lies, and levitical legerdemain have always been pretty much the same worldwide. (See pages 50–51.)

4

China: The Mind-Fist

"That which is decent in Japan is indecent in Rome, and what is fashionable in Paris is not so in Peking."
—Voltaire

The Chinese *Zizhi-dao* (pronounced dzee-gee-dow) means "the art of control." "Control" in this instance can refer to the authoritarian type of control—external control of our fortune and fate by gods and governments, but Zizhi-dao also applies to the amount of personal control we exert over our lives.

Gaining more personal control over our lives means taking more responsibility, disciplining and challenging ourselves both physically and mentally before our environment and/or our enemies do.

In the East, better control over your physical fate in general and your physical health and safety in particular is often accomplished by mastering one of the many martial arts.

In China, what we in the West call *"kung fu" is* called *"wu shu"* (war art). Kung fu (sometimes written *gung fu* or even *gong fu)* actually means both hard work and the excellence and mastery that grows out of that hard work.

Kung fu thus also applies to mental arts as well—both the mental control we exercise over ourselves, as well as the influence (cunning and force of will) we deploy to overshadow others.

To successfully overshadow another's mind—whether a military foe (through our use of propaganda and ephemeral strategy) a mugger on the street, or some authority trying to grill us concerning our whereabouts on the night in question, we use what the Chinese call *"gancui"* (pronounced gang-kway) which literally translated means "to penetrate neatly and completely." This refers to "getting inside the head" of our enemy.

However, part of the technique, getting inside an enemy's mind is not meant passively, as in trying to figure out where he's coming from. Gancui means *total mind penetration*—stabbing into our enemy's mind efficiently and completely, literally jamming disturbing and doubt-ridden thoughts, images, and paralyzing emotions like fear directly into his brain. Such a (counter) attack penetrates deeply into our foe's psyche, overrunning his defenses, stifling his attack, allowing us time to either finish him off with a scathing counterattack or else giving us time to escape the battlefield unscathed. No matter if we face our foe on a physical battlefield or a battlefield of the mind, the principles of victory remain the same. . . .

Total mind penetration means using our Mind-Fist.

Zizhi-dao, when applied specifically to mind penetration is also known as *I-Hsing,* the "Mind-Fist." I-Hsinq mind control should not be confused with the respected Chinese martial art of Hsinq-I.

Hsing-I kung fu (aka *Yueh Fei Ch'uan*) was founded by Chinese military hero Yueh Fei during the Sung Dynasty (960–1280 CE).

Hsing-I is a formidable fighting style of kung fu wu shu that emphasizes grace and economy of movement, with powerful straight line fist attacks.

Sun Yat Sen (Sun Wen) achitect of the 1911 revolution and first (and last) president of the short-lived Chinese Republic was a master of Hsing-I. And, judging by his masterful juggling of the volatile politics and the various fiery personalities contending for control of China during those explosive times, Sun Yat Sen must also have been an initiate if not an adept at the mental art of I-Hsing.

I-Hsing, the mental art of the Mind-Fist, is often studied hand in hand with Hsing-I, the more physical side of the equation. The balance of healthy mind/healthy body being respected both East and West.

FYI: An aid to remembering which is which: In "I-Hsing" (mind control) the "I" always comes first.

That's because, in real life, you have to look out for "number one"— yourself—first. Remember "the last can of beans" scenario? (See Lung, 2006b:31.)

If you don't look out for number one . . . people will "number-two" all over you!

Recall it was rumors of (and the reality of?) of Chinese *Hsi Hao* mind-manipulation techniques being used on Korean War prisoners that turned "brainwashing" into a household word in the West (Ibid., p. 50).

Down through China's colorful history, various cadre have used a vast array of mind control techniques—some to gain control over self, some to lord over others; some for the betterment of their fellows, some for the enslavement of their fellows. These tactics and techniques range from profound self-mastery secrets from ancient Taoist immortals, and tactics of intrigue and intimidation wielded by *Moshuh Nanren* "ninja," through masterful strategies from accomplished generals like Sun Tzu and Cao Cao, down to twentieth-century skullduggery and sundry slayings (mind and otherwise!) by both patriotic and profiteering secret societies.

As the blade itself is indifferent, we can only blame the hand for which direction it cuts.

Soon, with continued study, we too will be given the opportunity (and temptation!) to choose how we will use what we learn here today.

Choose wisely, Grasshopper!

MIND LIKE WATER

"The Tao that can be spoken of is not the true Tao."
—The *Tao Te Ching* of Lao Tzu

Confucianism, Buddhism, and Taoism are the big three in the history of Chinese religious thought, each having been influenced by the other two more than any of the three care to admit.

Confucius and Lao Tzu, the founder of Taoism, were contemporaries (sixth century BCE), although accounts and legends differ as to whether the two actually ever met.

Legend has Lao Tzu, the "Old Dragon," disappearing at the end of his life after penning his Tao te Ching. Some say he headed for India where he became the tutor for Prince Siddhartha, who was himself destined to become the "Buddha."

A thousand years later, the twenty-ninth Buddhist patriarch Bodhidharma (called Tamo by the Chinese and Daruma in Japan) took his particular (some say peculiar) brand of Buddhism from his native India to the Shaolin monastery near Foochow, China, where it blended with indigenous Taoism to create Zen Buddhism.

In the Taoist conception of the universe, everything exists within the Tao and the Tao exists with all things. In fact, there is nothing but Tao in all the universe. Even our thoughts when thinking about the nature of Tao is still Tao.

Things around us appear to be different from one another but this is an illusion because Tao is always the same—everywhere, always, and perfect.

In this way, Tao closely resembles the "Force" animating the movie *Star Wars*—all-powerful, yet impersonal, and available to the Luke Skywalker and the Darth Vader alike.

However, according to Taoism, we humans perceive things imperfectly, seeing our world as composed of warring opposites (hot-cold, light-darkness, etc.) when in fact there is only the one Tao. So instead of seeing Tao as a singular, unchangeable circle, we see it as divided, ever-changing "yin-yang."

We then go on to further divide up these yin-yang "opposites" depending on whether we perceive them as rising ("ascending") or in decline "descending"). This in turn gives rise to the five elements of which all things in the physical universe are composed (see figure 8).

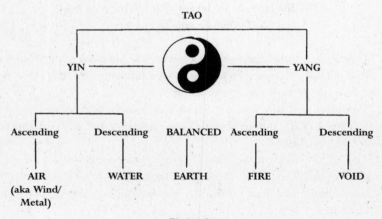

Figure 8.

72

While Confucianism is essentially down to earth, Taoism (pronounced "Dowism") is an airy-fairy philosophy which might have been popular with the modern beatnik or hippie. . . . Taoism was eventually corrupted from its earlier simplicity by the introduction of the magical side of philosophy. Its followers set out on a search for the elixir of immortality, and engaged in spiritism. And so Taoism quickly degenerated into what it is today—a polytheistic system of spiritualism, demonism and superstition. (*The World's Religions*, edited by Norman Anderson. England: 1950)

While most Westerners readily recognize the black and white "yin-yang" as the symbol for Taoism, the element Water is actually the symbol Taoist immortals (i.e., "Masters") used most often when attempting to explain transcendent Taoist concepts to minds still trapped by three dimensions.

Water is the ultimate shapeshifter. This is why it is often used as the ideal to strive for in the martial arts. Students are encouraged to give up stiff, harsh movements in order to "flow like water," adopting a more natural grace of movement that allows them to more easily blend with their opponent as opposed to always meeting force with force. This higher level of training is especially apparent in such martial arts *as Tai Chi* (*Dal Qi*) and *Aikido*.

Not surprising (otherwise you wouldn't find it in this book!), this idea of water as both ideal and metaphor has also be applied by Taoists to the mental dynamics of human beings.

And, since this Taoist philosophy applies to states of mind, it opens the way for the savvy Black Science adept to manipulate . . . uh, I mean *utilize* that information.

The Taoists have made this simple for us by classifying and comparing the three dominant mental states ("attitudes") to the three states of water: ice, fluid, and steam.

And while each of us is dominated by one of these states, we are free (to some extent) to use the other states freely—though, sans prompting, we tend to stay in our dominant—safe and familiar—state.

Ice, water in its fixed, solid form, appears to be unworkable. Yet ice can easily be sculpted to represent any form when subjected to the hands of an accomplished sculptor.

Human beings dominated by an "ice" attitude are conservative, fixed, and in extremis stubborn in their outlook and output.

On the positive side, they tend to be dependable. They are also predictable—a godsend in an enemy!

Such people aren't the sharpest tools in the box. Imagination isn't their strong suit. So when explaining things to the "Iceman" always stick to the K.I.S.S. principle: "Keep it simple, stupid!"

This type of personality (although secretly desiring guidance and direction, i.e., "reassurance") responds best when they think it's "their idea," rather than that they are being manipulated . . . oops! There's that word again. I mean . . . *directed* by others.

Ice is difficult to navigate but, with patience, it can be put to good use. So, too, ice-dominated people.

These types are what Sun Tzu referred to as "expendable agents," usually good for only one thing, dependable for only one mission and then only while under constant surveillance and/or supervision.

Consider a spear made of ice: It works just *as* well as a "real" spear, only to melt and eventually evaporate soon after. Expendable, but useful— perhaps vital—to the immediate need.

Ice-dominated people often wait too long, and thus miss out on timely opportunities. When enlisting the aid of an ice-bound ally, it may be necessary to literally "light a fire under his ass" just to get him to loosen up and get moving. Beware, though, ice heated too quickly turns to steam and disappears.

On the other hand, being slow to act is something we want to encourage in our enemies.

For enemies: reinforce his stubborness and eventually, like an ungainly glacier, he will collapse under his own weight. Think George W.

Water is either "resting" or "rushing."

Resting water is infinitely malleable, a shapeshifter, fitting itself to any situation, any container.

Resting water-dominated people make perfect actors, gigolos, spies, and assassins. They are patient, inwardly balanced, and liberal in attitude.

They are persistent, which always pays off in the end—the same way single drops of water in the infamous "Chinese water torture" steadily drip-drip-drip onto a victim's forehead ultimately make the victim "crack" . . . in one way or another!

Nature forms hard sedentary rock in this way. And remember this is also how stalactites and stalagmites are formed, from a few drips of water at a time. But once formed, both are hard as stone.

Machiavelli and Charles Manson come to mind when thinking of resting water personalities, as do other cult leaders and political survivors, chameleons capable of color shifts at a moment's notice.

Striking a balance, i.e., being able to shapeshift at will, as opposed to being "trapped" and vulnerable within a fixed form like ice, or doomed to simply fade into the background, like steam, gives us the advantage over those less adept when the order of the day is adapt.

"Rushing" water eventually overcomes everything in its way, either through raging force and impact, or else through persistent wearing down.

Those dominated by a rushing water personality make the best artists (Michelangelo) and guerilla fighters (Mao Tse Tung). That's because they don't take "no" for an answer, nor are they likely to bow under to critics.

Acclaimed for their critical powers of observation, these individuals often fail to turn the light of inquiry inwards and, if not self-vigilant, end up endangering themselves, become as impulsive as steam or, more disastrous still, as stubborn as ice. Voltaire comes to mind.

Steam is ephemeral, undecided and unfocused, yet carries within it the potential of becoming either fluid or ice.

Steam-dominated people are sometimes hot tempered and can explode under pressure. Stress is not their friend. Anger can be their downfall, but so can hesitation, as they have a short attention span. Thus they sometimes err on the side of taking too long to make a decision or else they make a decision on the spur of the moment without regard to future consequences.

It's hard to pin a steam person down—who is fast enough to catch steam in their fist? They can be innovative—often because they never took the time to learn what *isn't* possible in the first place. Edison for example. They are full of ideas, plans, but lack the focus and sometimes commitment to carry through their visions.

Negatively, the steam person is unfocused, and confused, undecided as to which way to go. Like the Iceman, this individual sometimes hesitates too long. Hesitation = death.

But unlike Iceman, Steam-boy doesn't hesitate out of an inability to process information at a timely rate, rather Steam-boy often has too many ideas floating around inside his head at one time. Short attention span, remember?

Leonardo da Vinci was a steam-boy. For all his genius, for every magnificent work of art and invention attributed to him, there are a dozen of his masterpieces left unfinished, scores of his inventions never built, left unrealized. For steam-boy, the ideas can come so fast he never actually gets around to completing any of them.

Steam-boys need somebody to help them coalesce and condense their ideas—be that somebody. Help him make his plans and make him stick to it.

It is a truism that a general must monitor a battle constantly, adjusting his plans at an instant's notice. A steam-dominated commander on the other hand will change his mind and subsequently his plans—arbitrarily—a dozen times until he loses the confidence of his own men.

Confronting a steam-dominated enemy, we encourage his confusion, his indecision. And we never forget that he is anger prone. (See page 77.)

We use these three attitudes in two main ways: Draw our enemy out of his dominant comfort zone, weakening and confusing him further before finally trapping him in unfamiliar territory and untenable ground where his dominant personality trait becomes the liability that ultimately dooms him. Or note the "weaknesses" inherent in each of these states, thus in our enemy, and reinforce and exploit that weakness to our advantage.*

THE TAO OF CAO CAO

"Nothing is constant in war save deception and cunning."
—Cao Cao

Cao Cao (also written Ts'ao Ts'ao) was the most successful general of China's Three Kingdoms Period. Living from 155 to 220 BCE, he rose from obscurity to eventually be crowned king of Wei, a kingdom controlling a sizable portion of China. His son went on to become emperor of all China, honoring his father as the true founder of the Wei Imperial Dynasty.

During his formative years, Cao Cao studied and benefited from those strategists who had come before him, most notably Sun Tzu.

---

*For more on Taoist mind magic see the "Tao You See It, Tao You Don't" chapter in Lung, *Mind Control*, 2006.

"THE THREE ATTITUDES OF WATER"

Shape	Essence	Attitude	Strategy to Use
ICE	Solid, fixed, appears unworkable.	Positive: Dependable, predictable. Negative: Unimaginative, unreasonable, stubborn in extremis. Cold & conservative toward others	Ally: K.I.S.S. when talking to him. Simple and straightforward. Enemy: Encourage his stubborness, encourage him to stay on the defensive.
WATER (YIN— STILL)	Potential, reactive to changes in "the weather" and circumstances.	Positive: Adaptable, patient Negative: Hesitant, overly cautious. Suspicious.	Ally: Negotiate with him. Keep him on a "need-to-know" basis. Enemy: Feed his paranoia.
WATER (YANG— MOVING)	Active, flowing, rushing.	Positive: Persistent. Negative: Impulsive, forceful. Liberal toward others. Tepid. Cautious but tolerant.	Negotiate with him while planning your Pearl Harbor. If he is attacking, use the Judo principle to make him overextend himself. As he rushes forward, funnel him into your container.
STEAM	Ephemeral	Positive: Curious, playful, innovative when focused. Negative: ADD—short attention span. Unfocused, undecided. Has trouble committing himself. Seeks form, easily led.	Ally: Help him to focus on your cause. Give him small concise missions, one at a time. Enemy: Encourage his confusion, fog, and fugue. Put him under pressure until he explodes from frustration.

Figure 9.

But Cao Cao was also influenced by Taoist concepts. For example, he understood the basic yin-yang nature of human motivation:

> It is human nature to advance toward gain while retreating from harm.*

Likewise, his understanding of yin-yang, opposites that complement and actually depend upon one another, helped him discover the all-important, intimate and intricate connection between recognizing opportunity and seizing victory:

> Mastery comes from opportunity. Opportunity is mastered through adapting to changing circumstance.

This harkens back to the symbol of yin-yang, two "fish" one black one white, swimming back and forth, around one another, the seed of the one within the other. (See figure 8.)

Applied to the battlefield, this Taoist concept allowed Cao Cao to turn the tables on many an enemy.

Appearing weak, he would sucker the emboldened enemy in, only to suddenly reveal his overwhelming strength. Or in a deficit situation, Cao Cao would confuse his foe with the unconventional and, as he put it, "Move beyond the regular rules," doing the unexpected, snatching victory from the jaws of defeat. Think Henry V rallying his exhausted troops for one final charge that will bring them victory, "Once more into the breech!"

But although he understood the Tao ("Way") of Taoism, he ultimately developed his own strategy and style, his own "tao."

In Taoism, there is the "big" Tao (the Universal, the all) and the "little" tao, our own individual way of stumbling through life. This includes our individual karma, as well as those actions we undertake while alive.

Tao with small "t" also refers to various arts, and crafts, and practices that can help us gain more mastery over self and thus gain a more full understanding (apprehension) of the great Universal Tao.

In Japan where Tao is called "Dō" (pronounced "dough"), it is said "The big 'Do' can be understood by practicing the little 'dō.'" In other words, diligent attention to our art and craft—for example a martial art—

---

*This is our basic "push-pull" equation. See figure 8.

will grant us the focus, clarity, and ultimate mastery to understand the big Dō, the ultimate reality.

This is why Zen masters tell us if we can understand but one thing fully, then we will understand all things fully.

Like Sun Tzu before him, Cao Cao preferred to avoid bloodshed. But when conflict was unavoidable and physical confrontation called for, Cao Cao was legendary for his ability to quickly and efficiently mobilize his forces, arriving at the battlefield of his choosing, rather than fight the enemy on ground they had chosen. (You never let the enemy choose the battleground.)

But even before ever setting out on such a campaign, for Cao Cao intelligence was always job one. Cao Cao realized early on that with the right kind of intelligence—both the innate and the gathered variety!—war could often be avoided altogether. Worst-case scenario and war became inevitable, the wrong slip of the tongue, the right whisper in the right ear, and precise intelligence would at least bring a war to a speedy—and satisfying—conclusion.

Cao Cao's battlefield principles, based on actual combat experiences, became the basis for the *"Iron Wall Kung-fu"* style he developed.

Iron Wall Kung-fu specializes in building an impenetrable "wall" of what to the untrained eye appear to be passive blocking techniques but what are in fact devastating counterattacks.

For example, rather than merely deflecting an attacker's punch with your elbow (aka your "short-wing"), you turn your elbow "block" into an elbow "strike" that cripples the attacking arm.

These same *"Iron Wall"* principles apply to Cao Cao's psychological warfare: Make the enemy think you are defending, when you are really counterattacking, make him think you are fleeing, when you are really leading him into an ambush.

This principle is of several general principles reflected in his many maxims; maxims we should study, not just in order to appreciate Cao Cao's wisdom, but also with an eye toward adapting and applying those principles in our own world:

Win Before You Begin

"Fight only as a last resort." (Cao Cao)

"War is not a personal vendetta. Balance gain and loss before beginning."
(Cao Cao)

"Plan well beforehand. Know your enemy and choose your lieutenants well.
Count your troops, measure the distances you must travel, and commit
the lay of the land to memory. Do all this before stepping one foot on
the road to war." (Cao Cao)

"A wise general seizes victory when first he drafts his troops. He does not
gather water at the well a second time." (Cao Cao)

In other words, measure twice, cut one. Can you say Vietnam? Iraq?

Amass men, materiel, and most of all intelligence before starting out on
any campaign. Likewise, go out of your way to deny your enemy these same
three.

The greatest of these three is always intelligence. If you have the choice
between stealing an assassin's knife or his road map, choose the latter.

Above all else, a wise general keeps his own counsel:

"Winning begins and ends with keeping dispositions hidden." (Cao Cao)

Use the Right Tool for the Job

From the emperor on down, everyone has his part to play in the coming
battle. However, it is the general's job to make sure he has the right man
attending to the right job at the right time.

It is also vital to keep specific spheres of influence separated. In some
instances, it is better for the right hand not to know what the left hand is
doing. Just so long as the general knows his right hand from his left.

This means, early on in any endeavor, deciding who's in charge, what the
chain of command is to be, and, most important, how to keep the lines of
communication open.

"Court and combat are two separate spheres. Combat cannot be won by
ritual nor by court etiquette." (Cao Cao)

Sun Tzu cautioned that a war cannot be run by an emperor sitting on a
throne a hundred miles away.

It's true we now live in the age of instant communication across vast distances, still Sun Tzu's principle must remain sacrosanct.

First, an emperor (or boss of any kind) must trust his general, just as that general must trust his troops.

Second, the general in the field must know how to think on his feet. And, when necessary, a general must have a pair big enough to make hard decisions in the field without the emperor's blessing (or second guessing). The wise general does this in order to (1) save lives and (2) seize opportunity by the throat.

Sometimes it is necessary to deny the emperor a battle in order to win him a war.

Micro-managing only strangles the creativity and enthusiasm of your troops. It makes them think you don't trust them. Don't put someone in charge unless you're sure of their abilities in the first place, no matter if you are playing on the ballfield or the battlefield. If you have to constantly be looking over their shoulder, then you've picked the wrong person for the job.

If you doubt any of your troops are up to the task at hand, either do the job yourself or else break the (complicated) task down into more manageable components crafted to the capabilities of your troops. War is not the time for training. Let each person know their place:

"Let each sphere be responsible for its officers. Let each officer be responsible to his sphere. Organize your regiments through the use of flags and gongs and symbols, and use loud drums to clearly signal attack and withdrawal." (Cao Cao)

As to Cao Cao's comment that "combat cannot be won by ritual," Plains Indians learned this the hard way. They liked to play a "war" game called "counting coup," where the idea wasn't to kill your enemy but to get close enough to whack him with a coup-stick or else—bravest of all—touch him with your hand.

Now while we may applaud this Native American ritual test of warrior courage, most Europeans these Indians tried counting coup on simply shot them dead.

That's not to say Europeans didn't have their own disastrous rituals when it came to combat. Dueling pistols at ten paces comes to mind. And British soldiers learned the hard way that a standing phalanx of lobsterbacks armed with inaccurate muskets just made too tempting a target for Colonial rebels armed with Pennsylvania rifles who knew how to hide behind trees.

Rituals can get you killed. Rituals have their place in the basilica, not the battlefield.

Rituals are simply habits decked out in habits. And habits are predictable. There's nothing better than a predictable enemy. Don't be one.

Generals (i.e., the boss, the Big Kahuna, commanders of all sorts) occupy the unenviable no-man's-land smack dab in between the emperor (i.e., the powers-that-be) and the grunts out there in the trenches.

Not only must the general interpret the (all-too-often contradictory and unrealistic) goals of the emperor, he must also find practical ways for deploying his troops without destroying his troops. Says Cao Cao:

> Without engaging the enemy in battle, the wise general accomplishes his goal with his force unscathed. He is victorious under Heaven because he has not sacrificed a single man.

Again Cao Cao repeats his supreme ideal, an ideal inherited from Sun Tzu: *Win battles with a minimum of bloodshed.*

Cao Cao held that achieving this goal pleased even the gods. Thus the wise general is not just victorious, he is "victorious under Heaven," i.e., his reluctance to wantonly slaughter pleases the gods, gods who value balance—Tao—above all else.

According to Cao Cao, the perfect general possesses five qualities: Wisdom, Integrity, Compassion, Courage, and Severity.*

The general must also survey the lay of the land, in order to take advantage of river fords, the protection (or obstacle) provided by mountains, and he must keep himself constantly updated on the weather.

In modern parlance this means knowing the territory you are venturing into—whether battlefield or business.

Is the market ready for your radically new idea? What obstacles—from critics to circumstance—might you expect to run up against? Can those potential stumbling blocks—people and predicament—be turned into stepping stones? Or do they need to be crushed altogether?

---

*For a more in depth discussion of Cao Cao's Five Virtues, see Lung, *Mind Control*, 2006: 212.

This also requires you to get to know the natives. Who among "the natives" can you expect will oppose you, who can you turn to your advantage via *"The Killer B's"?**

Once the battle is joined, once you set to sea, it matters not whether the prevailing wind is in your favor or against you, you must deal with it. (See Musashi's "Crossing at the Ford," in chapter 5.)

Or, as Cao Cao puts it:

> When all things operate in his favor, the wise general thinks about what can go wrong. When Fortune has turned her back on him, he schemes how best to get her attention again. Advantage alerts him to disadvantage. Disadvantage drives him towards advantage.

What Cao Cao is saying here is that an enemy shows you a victory (i.e., makes you think the battle's won), and suddenly the only thing on your mind is winning. Your men are still dying on the battlefield but you're distracted, already writing your victory speech in your head. And, of course, that's when your enemy drops the other shoe.

That's why, even when you are in an advantageous position, you continue to watch your back.

I know what you're thinking: The deal's done when the buyer signs the check. Au contraire. The deal's done when you cash the check and the money's in your hand. What? You never heard of a check bouncing?

Don't be in such a hurry to snatch proffered bait. If your enemy dangles a piece of fresh meat in front of you, let's hope you're a vegetarian.

You're right. This is the old "If it looks too good to be true, it probably is" warning.

Just remember that when somebody's offering somebody else an "It looks too good to be true" offer, it had better be you waving it back and forth under your enemy's snout.

It's important to remember that a truce is not a win. It's just a breathing space for your enemy to re-arm.

---

*Blind, Blood ties, Bribery and Blackmail, Brainwashing, Bullying, and Burying them! Lung, *Mind Control*, 2006:39.

And that's why you never—ever—negotiate with terrorists. Time always works to the smaller army's advantage. Every day your enemy is still above ground is a day you are in danger.

Even if you succeed in disarming him, and you think he's sitting peacefully in his home. Be assured he is spending that time sharpening his butter knife, trying to figure out how to set his shoes on fire, and teaching his sons and grandsons patience—"Your time will come, my sons. And then we will have our revenge!"

Your enemy doesn't *think* like you think . . . that's why he's your enemy.

Even though Cao Cao himself was heralded as a great general, he never let it go to his head; he never lost sight of the most important fact a general must remember:

> In the end, all victory depends on the skill of the single warrior.

The man in the field. The grunt. Ultimately your troops must have faith in you, his commander, even as you in turn must have faith in your commander-in-chief.

Whether a combat unit in Afghanistan or a cult in Antelope, Oregon, the troops must have faith in the chain of command—if only in the immediate officer over them. Read Shakespeare's *Henry V*.

Victory thus depends not on the desires of the emperor, not even on the strategy of the general, but on the willingness of the troops to carry out the general's commands.

A craftsman's only as good as his tools. A general is only as good as his men. Hannibal, Julius Caesar, Patton, all knew this secret.

Cao Cao warns us not to call troops up during the coldest part of winter, nor during the hottest part of the summer. Respect your troops. Let them know they can depend on you and you will always be able to depend on them.

Never sacrifice good men. And never let good men sacrifice themselves (as good men are wont to do). That's what Sun Tzu's "expendable agents" are for.

Bad dogs are always kept on a short leash. But even good dogs have to wear a collar. The difference is good dogs soon forget the collar they're wearing. Bad dogs never forget the leash.

Ultimately, for the commander of troops, it's lonely at the top:

"The troops can share in the joy of completion, but not in the travails of conception." (Cao Cao)

The More You Sweat in Times of Peace . . .

Practice doesn't necessarily make perfect, but it damn sure makes good sense. Make your training scenarios as realistic as possible.

Nine-Eleven was a wake-up call for many martial arts instructors. While sparring and forms practice still have their place, the ultimate training scenario is "How do I go through those two goons blocking that cockpit door in order to keep their third buddy from crashing this plane! Those bastards are standing between me and my loved ones, between me and the rest of my life!"

There's your realistic training scenario. There's your mind-set for devoting your time, effort, and perhaps the rest of your life to an endeavor.

(By the way, terrorists don't care if they die or not. Guess what? We don't care if *they* die either!)

And, for a little less intense application:

You practice your "Hire me/promote me" speech. You practice your pick-up lines, your posture, and hopefully some of your love-making technique!

Life requires practice. Practice the same way you should live your life— to the hilt! Confidence is contagious.

. . . The Less You Bleed in Times of War!

> *"While the enemy is still formulating his strategy,*
> *that is the time to attack."*
> —Cao Cao

Perfect the art of the "suckerpunch"—both the physical kind and the mental kind. What's that you say, "Suckerpunching someone, attacking unexpectedly, is dirty pool"? Oh yeah, thanks for reminding me: Hit him with the pool cue too!

The only dirty pool is the one you drown in. If you're not in the game to win, stay out of the way of those of us who are.

FYI: It is estimated that only 10 percent of animals are true predators, lions and tigers and such, hunting other animals for sustenance. If we extrapolate this to humans, that means that 10 percent of us are predators . . . and the rest of you are prey. Heh-heh-heh.

Hesitation = death. Indecision = death.

And when you war, war to the bone!

Cao Cao warred to the bone; he used a "double strategy" inspired by the Taoist yin-yang principle, always noting the opposite, always striving to balance the equation:

"The wise general attacks when his troops are twice those of his enemy." (Cao Cao)

When attacking, Cao Cao was again mindful to use his "two-headed snake" strategy, employing both regular troops and conventional direct (*Cheng*) formations, as well as unconventional and indirect (*Chi*) operatives. For example, he created "dragon-wings," specially trained and equipped shock troops (akin to modern special forces) that could travel great distances in short periods to either surprise the enemy (by arriving early at a battle site and laying in ambush) or else to relieve beleaguered allies.

Cao Cao also made liberal use of spies and, when prudent, *Lin Kuei* assassins. This "two-headed snake" strategy allowed Cao Cao to attack with direct attack on the one hand, surprise attack on the other.

Cao Cao always controlled the battlefield and never allowed an enemy to "draw him out" into open ground when his strategy was to defend fixed battlements. Likewise, he was quick to avoid and if need be break out of any restricting ground his enemy might try maneuvering him into, e.g., a boxed canyon, his back to a mountain, or to the sea:

"My enemy responds front, I attack his rear. When he responds by reinforcing his rear, his front again becomes vulnerable." (Cao Cao)

Responding to the enemy's every nuance was key to Cao Cao strategy—indeed, is key to any strategy!

Whether on the battlefield or in the boardroom (or the bedroom for that matter!), you must always be capable of instantly responding to shifting dynamics. Cao Cao speaks from experience:

Let the enemy create your victory.

How do we accomplish this? Cao Cao has an answer for that too:

Cultivate strength, spy out weakness.

Thus, anytime you make a move it is in response to a move made by your enemy. This doesn't mean we allow him to lead us around by the nose. It means you better be leading him around by the cojones!

The trick is to get him to respond to your prod or proffered opening, and then you, in turn, instantly respond to his predicted response. Yeah, it is something like chess. Thus:

"The wise general coagulates and dissolves his formations by first observing his enemy." (Cao Cao)

Pop quiz. Who figured out: "No battle plan survives first contact with the enemy"?

You're right! The correct answer is *every* great strategist since Sun Tzu. But, the Prussian Karl Von Clausewitz (1780–1831) gets the credit. (It really does pay to write your thoughts down!)

So your battleplan must adapt to circumstance and flux, evolving as it encounters obstacles and opportunities.

A scientist once told a fundie, "It doesn't matter if you believe in evolution. . . . Evolution believes in you!"

In other words, it doesn't matter whether you "believe" you need to adapt—"evolve"—in order to survive and better provide for kin and kind. You do need to adapt—constantly. And if you don't adapt, then you just don't . . . survive, that is!

Survival isn't for everybody. Just ask the dinosaurs.

Always have a "plan B." And for Black Science graduates, "plan B" stands for "*Be* ready to come up with plan C, and plan D, and . . ." well, you get the idea.

Never say "die", or you just might!

"Defend, when you lack strength. Attack, when your strength is in abundance." (Cao Cao)

"When the enemy attacks me, he becomes vulnerable. When he shows an

87

indication he is advancing, I set an ambush in place for him and feign retreat. If he seems anxious to retreat, I slow my march so he can do so." (Cao Cao)

"Attack where he is empty, avoid where he is full. The easy victory lies in subtleties, in attacking the vulnerable, not the invulnerable." (Cao Cao) "Attack what he holds dear and he will rush to the rescue and to his doom." (Cao Cao)

In other words, never take up the sword unless you know yourself capable of using it. And should you decide to fill your fist with steel, and steel your mind for what must be done, no matter how already keen their edges, remember to coat both with a liberal smearing of ruthlessness:

"When you look at me, don't see something you hate. See the thing you love the most. For that is what I will take from you if you rise against me!"
—The Hour of the Wolf

It Is Always Better to OutThink Rather Than to Have to OutFight the Other Guy

"Beyond the regular rules . . . One must use subtler origins."
—Cao Cao

RED SPEARS, BLACK HEARTS

"Wo xi wang ni man man si, dan kuai dian xia di yu!"
("I wish you a slow death, but a quick journey to Hell!")

From 1368 to 1644 China was ruled by the Ming Dynasty. In 1644, all that changed when China was conquered by the invading Manchu (also called the Ching). This invasion changed how everyone did business, including China's numerous secret societies.

According to Sterling Seagrave in his Black Science must-read *The Soong Dynasty* (1985):

There have always been such secret societies [in China], pirat-
ical bands, and esoteric sects. But until the seventeenth century, they were fragmented and iconoclastic. The year of the
Manchu conquest, 1644, was a watershed. After that a new
network of secret societies spread across the landscape, dedicated to unhorsing the Manchus. (p. 72)

The Manchus succeeded in subjugating northern China, but in the
south, stiff resistance sprang, rallying under the black flag of a pirate leader
named Coxinga. Outwardly, Coxinga threw his support behind the prince of
T'ang as Manchu emperor but, de facto, Coxinga ruled south China on both
land and sea.

Coxinga was a master strategist who came from a mixed Chinese-Japanese
heritage, his father an infamous Chinese pirate in his own right and his
mother a Japanese concubine.

When Manchus overran Nanking in east-central China in 1659, Coxinga
withdrew his forces to rugged Fukien Province and seized Taiwan (then called
Formosa) from the Dutch who were controlling it at the time. Unable to dislodge or otherwise defeat Coxinga, the Manchu finally succeeded in bribing
his father.

Despite this betrayal, Coxinga continued to rule Taiwan and a portion of
the mainland until his death in 1662—either from an epileptic seizure or by
way of poison.

Seagrave credits Coxinga's rebellion-resistance with giving birth to a new
age of secret societies (Seagrave, 1985:73).

The Rise and Fall of Shaolin

All who have ever written on the history of Asian martial arts always tip
their hats to the pivotal part played by the Shaolin (often written Shao-lin)
Order of monks.

As already alluded to in the section Mind Like Water, the twenty-eighth
Buddhist patriarch Bodhidharma is credited with founding the Shaolin
Order. Still much of Shaolin history remains a mystery.

The usual story is that Bodhidharma traveled from his native India
around 520 CE to what was then a Taoist monastery at Foochow, where,

after meditating for nine years staring at a wall, Bodhidharma had a revelation, out of which he formed Zen Buddhism.

What Bodhidharma taught was called *Dyana* ("meditation") in India.

In China this sitting meditation technique was called *Shan (Ch'an),* a word that means meditation but which is also a play on *Shan* ("mountain") because a monk sitting upright and still in meditation takes on the shape of a mountain.

In Japan, Ch'an is written "Zen." Zen therefore grew out of a fusion between indigenous Chinese Taoism and Indian Buddhism.

Bodhidharma, already the acknowledged twenty-eighth Buddhist Master, was now heralded as the first "Zen" Master.

Bodhidharma is also credited with teaching the monks at Shaolin a form of Indian martial arts designed to help strengthen them for staying awake during marathon meditation sessions. He is therefore honored in Asian martial arts circles *as* "the (grand)father of the martial arts."

He taught these "kung fu" exercises only for medicinal purposes . . . and never intended they should ever be used to kick serious ass. Or so the legend goes.

The fact of the matter is that—probably even before Bodhidharma showed up, and definitely after he shuffled off this mortal coil—the brothers of the Shaolin Order stayed knee-deep in controversy.

As was the custom in many Asian countries at the time, anytime a general got too old, or else when a general was disgraced and disciplined, or even when such a general came under suspicion of being just a wee-bit too ambitious, often such generals were ordered by the emperor to become monks.

So for every dedicated Shaolin brother seeking enlightenment, you could expect to discover another one stewing over being "exiled" to a monastery and seeking revenge. Not surprising then that the various Shaolin monasteries should become breeding grounds for sedition.

And it didn't help matters any when Bodhidharma introduced his devastatingly effective fighting art—originally known *as "Eighteen-Hands of Lo-han" (Lo-Han* being another name for "monks").

Soon some of the brothers at Shaolin were spending more time practicing their spear-hand thrusts than they were meditating on Buddhist sutras.

And, of course, every would-be rebel within a hundred miles was beating a path to the monastery gate, trying to spy out the secrets of this new fighting art. Predictably, before long, the "secrets" of Shaolin Kung Fu began to seep out, giving birth to a dozen other styles of fighting—and killing!

Over the course of years, several Shaolin monasteries, including the one at Foochow, fell out of imperial favor and were literally burned to the ground on orders of the reigning emperor—only to win favor again and be rebuilt by a subsequent emperor.

The story is told of how the Shaolin Order came to the rescue of the then-reigning emperor who then rewarded Shaolin with his blessing, commemorated by a three-sided imperial seal.

It is from this three-sided emblem that the dread Triads, the Chinese "mafia" would one day take its name.

Like everybody else in 1644, the Shaolin Order was greatly affected by the Manchu conquest (Seagrave, 1985):

> Coxinga's resistance movement included a powerful spy system and underground. After his death, his underground fragmented. But 128 militant Buddhist monks, who were part of the resistance, held out against the Manchu army at the Shaolin monastery near Foochow. They were unusually skilled at the martial art that is now called Kung Fu. A traitor betrayed them to the Manchus, and only 18 escaped.
>
> One by one, the heroic monks were tracked down until only 5 remained. These 5 Kung Fu masters became the nucleus of a new anti-Manchu resistance, which was organized along the lines of the ancient sects and pirate guilds of classical China. It was given the name Hung League after the first Ming Emperor, Hung Wu. (p. 73)

The Hung League's rally cry was *"Fan ch'ing fu Ming,"* "Down with the Ching, up with the Ming!"

Seagrave thus credits the Shaolin Order of "peaceful" Buddhist monks with giving birth to damn near every cut-throat Chinese secret society that was to follow, including the Chinese "mafia"!

But it was not just fighting techniques (*wu shu quan*) that filtered out of Shaolin, eventually filling the fists of less spiritually minded individuals. Tactics and techniques of mind control—both over self and others, called Jing

Gong, also escaped over the walls of the Shaolin monastery, only to be eagerly embraced by rebels fighting against the Manchu, and by every shadowy secret society with a bone to pick—or break!

Chinese Secret Societies

> *"Silence was both a game and a necessity.*
> *Punishment for a loose tongue meant its removal."*
> —Seagrave, 1985:64

As already mentioned, secret societies had a long tradition in China, long before the Manchu conquest of 1644.

As with secret societies in the West, Chinese secret societies can be divided into two types: religious-motivated groups, called *Chiao*, and the more political—and often outright criminal—motivated secret societies known as *Hui*.

As can be expected, there was much interchange between these two types. Religious-motivated secret societies often dabble and delve into politics and/or degenerate into criminal activities. And, as then as today, as there as here, there have always been criminals who find it convenient to hide behind a priestly cloak—dagger optional.

For example, in the pre-Manchu period, the most active secret society, the White Lotus Society, straddled the fence between religion and politics, having been described *as* "originally a militant, messianic Buddhist order" (Hucker, 1975).

After the Manchu invasion secret society activity increased and unrest intensified, especially following the death of Coxinga in 1662.

Throughout the 1700s, up into the first half of the nineteenth century, uprisings continued.

The infamous Opium War (1839–1842), where the mighty Manchu were forced to kowtow to Western powers, showed the vulnerability of the Manchu Dynasty to the Chinese people.

Various secret societies and subversive groups took heart from this and took advantage, seizing every opportunity to provoke trouble between the Manchus and the Westerners. The enemy of my enemy is my friend.

Unrest continued to grow, culminating in the horrific Taiping Rebellion of 1850–1865.

FYI: The 1850 *Taiping Rebellion* has been called the "most destructive civil war of world history" and is credited with heralding "the end of China's traditional history and the beginning of modern history" (Hucker, 1975:268).

The Taiping Rebellion started as a popular uprising of "Taipings," fundamentalist Christians led by a south China "messiah" claiming to be the younger brother of Jesus Christ. Their name came from T'aiping T'ien-kuo, "The Great Peace Army of the Heavenly Kingdom." In their minds the battle was between the Heaven-guided Taipings and what they saw as the "godless" Manchu.

The Taiping Rebellion raged for fourteen years, across twenty provinces, leaving over 20,000,000 Chinese dead.

Unable to put down the Taipings on their own, the Manchu finally hired a mercenary army of Westerners dubbed the "Ever-Victorious Army." This EVA quickly broke the back of the Taipings when the traditional way of Chinese fighting (clustered up in a mob) proved no match for modern Western military tactics.

In the end, the younger brother of Jesus committed suicide and the rebellion was crushed.

Several important changes came out of the Taiping Rebellion.

By admitting their own inability to crush the rebellion, and by having had to resort to hiring Western mercenary troops to crush the rebellion, the Manchu "lost face."

For the first time, Chinese peasants saw that, with the right leader, they could challenge the ruling Manchu.

Equally important, for what was to come later, in the aftermath of the failed Taiping Rebellion, many triads who had thrown their considerable weight to the anti-Manchu rebellion gave up on politics altogether, turning their considerable talent for cunning and craft instead to simply making a profit. Some of these triads would become major players during China's revolutionary period at the dawn of the twentieth century.

The main thing to come out of the Taiping Rebellion was that the Chinese people now hated both the Manchu and the Westerners equally.

Widespread flooding and drought in 1898 and '99 led to more discontent among the Chinese people, which, in turn, led to another surge in secret society activity, setting the stage for the next great uprising.

Whereas the Taiping Rebellion had been aimed at the "godless" Manchu, ironically, the next great uprising would ultimately receive the blessing of the Manchu.

The 1900 Boxer Rebellion set out to cleanse China of foreigners, especially Christians.

The Boxers, aka "Fists of Harmony and Justice," were a bizarre blend of *Chiao* and *Hui*, freely mixing Shaolin boxing (hence their nickname) with various Taoist and Buddhist mind-strengthening exercises and, just to hedge their bets, they also used mystical shamanistic incantations and spells designed to make them invulnerable to Western weapons.

Boxers identified themselves by wearing a red sash, symbol of a martial arts master who has killed an opponent in combat.

Unlike the mostly southern-based Taiping Rebellion, the Boxer Rebellion had its strongest support in the Manchu-dominated north and would eventually be co-opted by the Manchu empress to target foreigners.

Before it was over, Boxers murdered 200 foreign missionaries along with fifty of their children, and slaughtered over 20,000,000 of their countrymen who had converted to Christianity. They also besieged the special foreigners' settlement in Peking (nowadays called Beijing) for fifty-five days until a multinational force from Western nations came to the rescue. See Charlton Heston's marvelous movie, *55 Days At Peking* (1963).

The Boxer Rebellion fell apart after its leader Li Wen-ch'ing was executed. But the outrage it had unleashed against foreigners, not to mention the fact that the Manchu had backed a losing team, spelled the beginning of the end of the Manchu Dynasty, condemning China to the next fifty more years of revolution and civil war.

As China stumbled forward into the twentieth century it also ran headlong into civil war and chaos, as a myriad of warring factions and wide-eyed fanatics—political adventurers, private armies, and paid assassins—battled, bargained, and back-stabbed their way toward power, all scrambling to fill the void left by the terminally ill Manchu Dynasty.

But not only were there Chinese forces within China fighting, there were also hungry, sinister special interest groups outside China, licking lips in anticipation, determined to get their piece of the China pie.

Some of these shadowy foreign interests were "invited" into China by various Chinese political parties willing to literally make a deal with the foreign devil in order to get the upper hand against their rivals.

For example, Mao's Communists had already sold the souls they didn't believe in to the Bolsheviks.

Even the man fated to become China's first and only president, Dr. Sun Yat Sen, sought financial backing from both wealthy Americans and overseas Chinese criminal Triads (called Tongs in the West).

But Sun Yat Sen didn't stop there. In July 1897 he met with a group of what author Sterling Seagrave calls "powerful Japanese political ronin"—the founders of Japan's infamous Black Dragon Society, a secret society even then already hard at work laying the groundwork for Japan's eventual invasion and conquest of the mainland (Seagrave, 1985:84).

Of course, as always, secret societies and Triads within China were also hard at work. Many of these forces grudgingly came together under the banner of Sun Yat Sen's Alliance Party, backing the Revolution of 1911.

Sun Yat Sen's Alliance Party fell apart after his death in 1925, with Mao's Communists literally going one way and Chiang Kai-Shek's Nationalists going the other. This severance ushered in another twenty-five years of slaughter and skullduggery across the length and breadth of the Chinese mainland.

During these troubled—and for some, profitable!—times, secret societies maneuvered behind the scenes, each with their own agenda.

In many ways Mao's Communist Party was a giant secret society, at least in the early years when they were still the hunted, as opposed to the merciless hunters they later became!

Not surprising, during this time Chinese Communists did their best to infiltrate and subvert existing secret societies. Tit for tat, Triads and other secret societies were doing the same, trying to co-opt the Communist Party apparatus and avenues to expand their criminal ventures.

Recall that Sun Yat Sen himself was a card-carrying member of the *San-ho-hui* ("Three Harmonies Society").

Generalissimo Chiang Kai-Shek belonged to the infamous Green Gang Triad, who supplied Chiang with money, men, and materiel, all with Green Gang strings attached.

The Green Gang Triad eventually came to dominate all other triads in mainland China and corner the market on opium and heroin trade in China.

By 1937 the Green Gang—via their triad "tongs" in America—were supplying drugs to Lucky Luciano's La Cosa Nostra.

Even during the worst fighting in World War II when even the Chinese Nationalists and Chinese Communists had grudgingly put aside their internecine slaughter in order to combat invading Japanese, the Green Gang still continued to sell drugs to Japan uninterrupted. One dirty hand washes the other. In exchange for permitting Green Gang's underworld activities (including smuggling heroin) to operate uninterrupted inside Japanese-controlled territory, the Green Gang guaranteed the security of Japanese garrisons and businesses in the Yangtze Valley (Seagrave, 1985:373).

Seeing the potential of secretive agents operating behind the Black Curtain, Chaing Kai-Shek formed his own secret society, "The Blue Shirts," a combination secret society/Gestapo capable of dealing silently and suddenly with his enemies. Of course, Chiang was wily enough not to sever his ties with the all-powerful Green Gang.

As already alluded to, the Green Gang eventually displaced all the triads and secret societies in territory under Chiang's control except for a secret brotherhood known as the Society of Elders and Brothers.

One reason for this oversight was the fact that the Green Gang's power base was concentrated in major cities, whereas "field triads" like the Society of Elders and Brothers were primarily rural movements.

While what was going on in the major cities rated the front page of the newspapers, out in the countryside triads and other secret societies had also been formed, some out of necessity and self-protection, some, like their city cousins, purely for personal gain. The Society of Brothers and Elders was the strongest of these, the seeds of which had been planted with the 1644 Manchu conquest, and subsequently watered—and on occasion ruthlessly pruned!—by Coxinga's revolt, the Taiping Rebellion, and the Boxer Rebellion (Seagrave, 1985):

> Because the Manchu conquest had fragmented Chinese life, the countryside for decades afterward was full of drifters, farmer soldiers, vagrants, beggars, thieves, desperadoes, murderers, journeymen, tradesmen, businessmen, itinerant craftsmen, artisans, students, and political exiles. To accommodate such diverse and growing membership, the triads multiplied until by the nineteenth century there were hundreds of sec-

ondary guilds that maintained only symbolic links to the parent orders. Some, like the widespread Society of Elders and Brothers, were Robin Hood bands of poor people, farmers, and canal boatmen, who built a complete underworld economy on smuggling. Many of the Communist guerrillas of the twentieth century came from this organization. . . . The triads attracted adventurers who found the upward path blocked in normal society. Through them, ambitious men could control secret funds and idle manpower, gaining extraordinary political and social leverage. Some triads were strictly patriotic, while others used patriotism to disguise purely criminal pursuits; still others combined both purposes, with an invisible "dark thread" criminal organization, on which was superimposed a patriotic and visible "light thread." (pp. 74–75)

The Society of Elders and Brothers was descended though a distinguished (and disturbing?) lineage of thought and threat that included every art and artifice, craft and cunning from shamanisthus to Shaolin.

It is from the Society of Elders and Brothers that the Red Spears emerged. (See figure 10, next page.)

The Red Spears

"He used them as an instrument, playing upon their obvious
fanaticism, string by string, as a player upon an Eastern harp, and
all the time weaving harmonies to suit some giant, incredible
scheme of his own—a scheme over and beyond any of which they
had dreamed, in the fruition whereof they had no part—of the true
nature and composition of which they had no comprehension."
—*The Hand of Fu Manchu*

Some saw the Chinese Red Spears secret society as simply political terrorists (Laqueur, 1977). Others saw them as Chinese resistance fighters (Tuchman, 1970):

There were bands of resistance fighters drawn from the country people who, made desperate by marauding soldiers, had

SHAOLIN ORDER
(Foochow/Henan, etc. f. sixth century CE)

THE HUNG LEAGUE
(f. 1644)

"Light thread"

CHIAO
(religion-motivated)
[Concentrated in North]

a. SOCIETY OF ELDERS &
   BROTHERS
b. 8-TRIGRAMS SECT (aka
   "Fists of Harmny & Justice"
   bka "The Boxers" 1900)

THE RED SPEARS (c. 1900s)

"Dark thread"

HUI
(political/criminal motivated)
[concentrated in South]

TRIADS
("Chinese Mafia")

CHINESE CRIMINALS &
SECRET SOCIETIES

CHINESE-AMERICAN CRIMINAL
SECRET SOCIETIES
(aka "Tongs")

T'IEN-TI HUI
("Heaven & Earth
Society")

SAN-TIEN HUI
("Three Dots
Society")

SAN-HU-HUI
("Three Harmonies
Society")

Dr. Sun Yat Sen
(member)

BLACK DRAGON SOCIETY
(Japan, cf.
Seagrave, 1985:83)

CAO DAI religion
(Vietnam, Sun Yat
Sen made a "saint".
Visited Saigon 1907)

Figure 10.

organized the Red Spear Society to prey upon whatever small groups of soldiers they could handle. They killed without mercy, inflicting wounds that left their victims alive for three or four hours before they died. (p. 109)*

Still others have cited them as the inspiration for the *Si-Fan*, the "White Peacock," the nefarious criminal league in Sax Rohmer's acclaimed *Fu Manchu* series of novels that were forever warning us of the dread rising of "the Yellow Peril."†

When discussing the Red Spears, everybody gets it a little right. Perhaps that's the best that can be hoped for from outsiders looking in, given only smoke with which to blow off the dust from a mirror perhaps centuries old. For while the origins of the Red Spears is most often traced back to after the 1644 Manchu invasion, it is important to remember that the Red Spears "inherited" their tactics and techniques, both physical and mental disciplines, from a bevy of secret societies, sects, and cults that had gone before.

Some of these techniques were decidedly treacherous and undoubtedly murderous. And while we are here to glean what we can of the Red Spears' mind-manipulation tactics and techniques, we should never forget that the times the Red Spears lived—and thrived—in demanded physical survival first and foremost. And that often came down to simply bleeding less than your enemies!

But we can't paint all the Red Spears with this same broad—bloody!—brush. Over the years many drastically different cadre operated under the flag of the Red Spears, some actually associated with the Red Spears, some unintentionally or deliberately identified with them, and some gangs who

---

*One "dark thread" of the Red Spears developed an especially nasty technique called Bo Pi ("flesh peeling"). See Theatre of Hell: Dr. Lung's Complete Guide to Torture, 2003.

†The Red Spears, which some have dubbed a "cult," helped inspire author Sax Rohmer (real name A. S. Ward) to craft his popular *Fu Manchu* series of books. In turn, Rohmer's Fu Manchu novels, in particular *The Island of Fu Manchu,* gets the (dis)credit of having "inspired" Nation of Islam cult founder Elijah Poole (aka Elijah Muhammad) with concocting the bizarre racist clap-trap still used by his cult today, i.e., that the blue-eyed devil white race is a result of genetic engineering carried out on an isolated island by a "big head" mad scientist! (Stephen Howe, *Afrocentrism: Mythical Pasts and Imagined Homes.* London: Verso, 1998:72/fn5). First, this is a curious case of "Six degrees of separation," if ever there was one. Second, proof of the Black Science adage: "Cult begets cult."

simply usurped the banner as their own purely for the propaganda and profit they could wring from it.

Different "Red Spears" cadre were thus motivated by different goals, often far removed from the central Red Spear goal of benevolence (if only toward its own members).

It is said even those Red Spears "fueled by Chiao" were all too often "fanned by Hui." An acknowledgment how even the best of intentions—no matter how high the initial calling—can all too easily be manipulated by those with a hidden agenda and a knowledge of mind control to make it happen. Like Black Science graduates (heh-heh-heh).

The Red Spears were thus many different groups, held together by a loose ideology of "provision of brothers, paranoia of others"; a sliding scale of morality with some leaning toward the spiritual (chiao), some enamored by the revolutionary (hui), and some more concerned with the financial.

Like all secret societies in China, no matter their immediate orientation, Red Spears at least paid lip service to having been descended from Shaolin.

Even before the Shaolin Order was scattered to the four winds, Shaolin "secrets" had already begun spilling out into the world at large. The direction these teachings took once outside the walls of the monastery depended entirely on what band of rebels, ruffians, and out-right highway robbers they fell in with.

Originally, Shaolin teachings and training were divided into several levels known as "halls." Aside from fantastic claims the Shaolin Order had as many as eighty-one such training levels, most historians agree that this Brotherhood of Bodhidharma organized their knowledge into nine distinct training Halls, which, while unique in their own right, overlapped and complemented one another.*

We now know that these "Nine Halls" were actually "Three Halls," albeit with three levels of accomplishment each.

These three Halls were:

---

*After the dispersion of Shaolin, many of the shadowy scavengers picking up the pieces and picking over the scraps adopted this Nine-Halls teaching structure; some because it bolstered their spurious claims to being the "true" inheritors of the Shaolin mysteries, others, simply because it was a convenient and effective teaching structure. Most notable (or is that "notorious"?) among those who freely adopted this system was the Shinobi Ninja of Japan. (See *Nine Halls of Death*. Citadel, 2007.)

Ying Gong

Ying Gong promoted body toughening; calisthenics and yoga-like exercises designed to strengthen the body. These were those exercises originally taught by Bodhidharma in order to help the Shaolin brothers stay awake during marathon meditation sessions.

The focus was/is on strengthening the body outwardly, in order that the student could withstand the rigors of monastery life and/or survive in a war situation when called upon to do so.

Chi Gong

"Internal training" strengthened the internal organs and helped increase the student's overall "balance," both inside and out. These exercises purified and cleansed the internal organs and systems of the body, increasing overall health.

Moreover, Chi Gong also helped the practitioner reduce stress, relax, and cultivate "chi" energy. Once a student had mastered the basics of Chi Gong, they graduated to *Dai Qi Quan* (tai chi) exercises—taught first as calming and balancing exercises, only later at a more advanced level was the more mature student shown the devastating martial arts applications.

Jing Gong

"Sense training" is designed to increase a student's overall awareness through teaching them to use their five senses fully. At this level, the student's previous forging of a strong body-mind awareness is expanded upon. As a result, the student's adroitness of mind not just doubles, but expands exponentially as the student masters first his own mind, and then develops the power to master the minds of others.

In keeping with their Taoist roots, Red Spears concentrated on the two main Shaolin disciplines: Ying Gong and Jing Gong, body and mind, a balancing of yin and yang.

These two disciplines were practiced to varying degrees by different cadre depending on individual mind-set and collective motivation.

For the more military minded, for practical reasons, there was an emphasis on Ying Gong—body toughening and Shaolin fighting technique.

Others, the more chiao-oriented, preferred to follow the wise Buddha's Chi Gong observation that "Your best weapon is in your enemy's mind."

Red Spears inherited much from Shaolin, not just Wu Shu fighting arts, but also inner disciplines like Jing Gong. As a result, even those who were in awe (fear) of the Red Spears capacity (some would say propensity) for swift and efficient violence, nonetheless also took note of their special disciplines, if not their "powers". According to Walter Laqueur (1977):

> The "Red Spears" of the 1920s combined politics with exercises such as deep breathing and magic formulas, rather like the counterculture of the 1960s. But politics was only one of their many activities and in this respect they resemble more the Mafia than modern political terrorist movements. (p.9)

Did you note the reference to "deep breathing" (obviously Ying Gong) and to "magic formulas"? More on those magic formulas in a minute.

All secret societies try to wrap themselves in a cloak of mystery. This is done to discourage scrutiny and for protection—reputation spills less blood!

Like many shadowy groups, Red Spears were credited with possessing mystical powers. Some of this credit came from outsiders in awe of the Red Spears' abilities, while other such credit was pure propaganda and self-promotion on the part of less scrupulous Red Spear cadre.

On one hand, cultivating superstitious awe and promising to bestow supernatural powers on the worthy was/is a marvelous cult recruiting tool. Still today, cults promise to teach recruits how to "open your third eye" and "unleash your hidden mystic powers." (Hucksters in the West are no strangers to this "promise of powers" ploy; from Rasputin down to Charlie Manson, and every one of those lames today with claims to be able to talk to the dead.)

In the East, shadow-knights like the Assassins, India's Thuggee, China's Lin Kuei, and Japan's Ninja all knew the advantage of playing on their enemies' superstitions, of feeding their foes malarkey about their possessing magical powers to fly, to walk through walls, and of being able to kill with "the death touch" (see Lung, 1998).

The Boxers used dazzling displays of Shaolin martial arts, sleight-of-hand magic, and techniques of mind manipulation to capture the attention of, and then pull in, peasants to their cause.

Ying Gong Senses Training

"All that I have to say has already crossed your mind."
—Professor Moriarty

"Then perhaps my answer has already crossed yours."
—Sherlock Holmes

With self-mastery of your own mind soon comes (the temptation of) mastering the minds of others.

Some Red Spears practiced varying degrees of mysticism and mind control; other Red Spears leaned more toward the political; still others leaned toward criminal endeavors.

There are thin lines between magic, mysticism, and true mind control. Whereas the first two can all too easily be faked, true mind control requires first exorcising our own demons before learning to exorcise the demons inside our enemy's head!

For example, hypnotism was once purely the domain of sundry performers; however, it was soon co-opted and corrupted by panderers and perpetrators of fraud until it became more legitimized in the nineteenth century for use in medicine and psychiatry.

Like many who find their succor in the shadows, who fan and feast on the fears of their fellows, Red Spears realized that if their enemies believed they possessed a "sixth sense"—granting them the gift of mind reading and mind control, this belief created an added layer of hesitation between them and their enemies.*

Developing the full use of our five given senses does often make it appear to the more indolent, ignorant, and superstitious-prone of our peers that we possess some magical "sixth sense." ESP, if you will.

Thus the Red Spear motto: "Sow five, harvest six," meaning that mastery of our five given senses might actually grant us a sixth sense.

And making your enemies think you have special powers keeps them guessing and is always good strategy. It never hurts if your enemies think you have mystical powers . . . unless you live in Salem, Mass.!

---

*Remember, you can't write "believe" or "belief" without writing a "lie" in the middle.

That some groups of Red Spears dabbled in, or even dealt exclusively with, mysticism, whether out of true belief or else just to confuse their enemies, is well documented.

But we need not get brain strain when seeking out the sources of Red Spear mysticism and mind control.

Starting out as a primarily rural movement, the Red Spears were influenced by ancient Chinese shamanism, much of which was eventually incorporated into Taoism. This indigenous shamanism included faith healers, fortune-tellers, and mountain mystics of every ilk.

Later Buddhism would arrive to put new icing on this already multilayered cake. By the time these shamanistic, Taoist, and Buddhist techniques had made their way to the Red Spears—by way of Shaolin, by way of the Society of Brothers & Elders, and by who knows what other secret society filters—it was already a heady concoction distilled from a hundred heavily laden vines of thought.

Some of these techniques complemented and expounded upon one another. Other times the latter tried to usurp the truth of the former, leading to the truth of the former ultimately betraying the false intentions of the latter. But, once the dust settled, all proved useful—thereby surviving the cut, winning the honor of then being passed down while other less worthy, less time-tested tactics and techniques withered by the wayside.

The truth is in the doing.

By their very secretive nature, many of the Red Spear esoteric rituals and techniques for mind control and manipulation have been lost. Fortunately, for our purposes, others have been preserved.

We know, for example, that many Red Spears paid homage to the myth (?) of "The Nine Unknown Men" (see Buddhist Black Science, in chapter 3).

We also know that the Red Spears received much of their knowledge through the Society of Brothers and Elders who, in turn, received much of their inspiration—if not initiation—from Shaolin, by way of the original Hung Society.

Pat Robertson, in his *The New World Order* (1991) credits (accuses?) the Hung Society—and though not mentioning them by name, Shaolin—for helping inspire Western Freemasonry:

> The candidate [for Masonic initiation] therefore must strike
> back at those assassins which are, courtesy of the Illuminati,

the government, organized religion, and private property. . . . This particular ritual is not Egyptian but from the Hung society of China, based on the cult of Amitabha Buddha. The ceremony which clearly resembles those of the *Egyptian Book of the Dead* was apparently copied by the Freemasons. It involves not a builder named Hiram, but a group of Buddhist monks, all but five of whom were slain by three villains, one of whom was the Manchu Emperor Khang Hsi. (p. 185)*

Yeah, when you get right down to it, people *are* pretty gullible. Hey, Dr. Lung just *reports* the news, okay?

So if you weren't born with ESP, or haven't yet mastered those Jedi mind tricks . . . fake it till you can make it!

Use every trick in the book (especially in *this* book!) to convince your enemy you possess abilities to crush him, up to and including the supernatural.

But in the end, nothing is better than actually devoting the time and attention it takes to actually develop your own mind. Begin by making your initial goal the full mastering of your five senses.

And if along the way you inexplicably discover yourself with the ability to "hear" other peoples' thoughts, to bend them to your bidding through your mesmerizing gaze alone, and if objects suddenly start levitating around the room . . . well, I'm sure you'd never use such special mind powers for selfish, evil, twisted gain (heh-heh-heh).

So far twenty-first-century scientific discoveries have only served to prove what ancient Taoist masters, Shaolin priests, and Red Spear adepts intuited about the potential—the powers!—to be gleaned by giving our five senses our full attention.

---

*Freemasonry repaid Asia in the 1920s when French Freemasons helped found the Cao Dai religion of Vietnam. FYI: Sun Yat Sen is canonized in the Cao Dai church. (See Lung, 2006a.)

Jing Gong Senses Training

"There are even senses that are never used."
—Voltaire

It's a simple formula: *Sensation* leads to *(Re)Cognition*, which in turn leads to *Thought and Action*.

The trouble is, smack dab in between what our senses encounter (sensations) and what our brain actually takes in (recognizes and decides to think about, are often *Physical Diseases or Defects* that prevent this second stage from taking place (e.g. hearing loss, varying degrees of blindness). (See figure 11.)

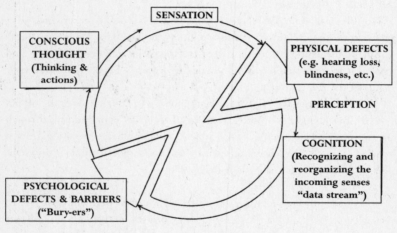

Figure 11.

Then, to make us question ourselves even more, in between our perceiving and recognizing a stimulus (a sensation) and our subsequent conscious response (thoughts and actions), all kinds of *Psychological Detours and Defects* can interfere with what messages our brain sends to the rest of our body. Our brain can even send greatly skewed information (misperceptions), either paralyzing us with fear or else kicking in our flight or fight reaction.

If you've read *Mind Manipulation* (2002) and *Mind Control* (2006), or if you've at least managed to stay awake during the first half of *this* book(!), then you know you can't trust your senses.

But all is not lost. You can, with due diligence, gain more mastery—perhaps even complete mastery!—over your senses. At the very least, you should aim to make your senses your servants, rather than remaining a slave to them.

According to Dr. Anthony Zaffuto, author of *Alphagenics: How to Use Your Brain Waves to Improve Your Life* (Doubleday & Co., 1974), the key to gaining full control (i.e., full awareness) of your senses is to, first, close off all sensory imput, and then, second, to focus and concentrate all your attention on one sense at a time. (Zaffuto, 1974:31)

This is the same method that's been used for centuries by mind masters and mystics both East and West. It is the basis for all mind-strengthening (and mind control) disciplines, including meditation and self-hypnosis, both of which were taught to students, first at Shaolin, and then by Red Spear cadre as part of their overall jing gong sense training.

In order for us today in the West to gain better control of our senses, it is first important to understand that, despite popular belief and pop psychology to the contrary, there's no such thing as "multi-tasking." Sorry to burst your bubble but your mind can only consciously concentrate on one thing at a time. The only reason we imagine we are thinking about and doing more than one thing at a time is because the human mind jumps from one senses-inspired thought to another like a two-year-old at Toys-R-Us.

The goal is to train (program) our mind to switch back and forth, and to race along the myriad of connections within our brain faster—and with more direction and purpose. In other words, to process incoming sense information in as effortless, expedient, and efficient a manner as possible.

This can be done. The mind can be trained. And we begin training the mind by first training the five senses.

Why bother to "master" your senses? First, surviving—for protection. The more aware you are of your surroundings, the better your odds of outwitting and, if need be, outfighting your enemies should (when!) society and civilization go horribly wrong . . . again!

And the only way to be more aware of your surroundings is to increase the attention you give to your senses since, duh!, it is only through your senses—seeing, hearing, smelling, tasting, and touching, that you interact with, and receive vital information from, your environment.

Sun Tzu's "Know yourself, know your enemy, and know your environment" can all be satisfied by our paying more attention to mastering our five senses.

Mastering our senses means we've successfully set out on the road to "knowing ourselves."

Mastering our senses, knowing how they work—or how they conspire to trick us!—means we know how our enemy's senses function as well; information we can use against him.

And, as just mentioned, mastering our senses makes us more "at one" with our surroundings—upping our odds of finding shelter, weapons, and food when the time comes.

For Red Spears, and other battle-beleaguered groups, being more aware of their environment literally meant the difference between life and death.

Second, beyond merely surviving comes thriving. A fuller use of your senses enriches your life.

Like a man born blind who has never seen a sunrise, you can't really appreciate the world around you unless you're firing on all cylinders, using the full potential of your senses.

But if survivin'-n-thrivin' ain't good enough reasons for mastering your senses, what if I told you it would improve your sex life?

Hindu and Buddhist Left-Hand Tantra disciples (recall we made their acquaintance in chapter 3 in Six Senses, Five Virgins?) practice sensory awareness exercises that allow them to direct their Prana (breath/vital energy) to different parts of their bodies. This is similar to the way a Chinese chi master can direct his chi-flow simply by willpower alone. This ability to direct prana (or chi) flow greatly enhances any experience—especially sexual pleasure.

For most men—anywhere!—sex is a "dick-thing," in other words, their orgasm is 99 percent centered in their penis. However, recent scientific discoveries have determined that, since all physical sensation is first channeled (filtered) through the brain, *all orgasms take place in the brain*.

Ancient Tantric masters already knew this, that by learning to control the flow of your consciousness (i.e., focus and concentration), you can have an orgasm centered in *any* part of your body. Imagine the possibilities.

By the way, focus and concentration are not synonymous. Focus refers to our "tightening" one of our senses (or our overall mind), sharpening it to a pinpoint directed at a singular, particular thought or object.

Concentration means holding that captured thought or object with our mind and focused sense for some period of time. Focus is how we stab the blade . . . concentration is how we twist it!

An increase in our overall focus and concentration is both the way in which we strengthen our senses, as well as the by-product of doing so. Thus, mastering our five senses literally begins and ends with focus and concentration.

But before we begin, review the section Six Senses, Five Virgins, in chapter 3 and take another look at page 29 and figure 6, on page 33.

Now You See It, Now You Don't

> **"Eyes given to see are not always open."**
> **—Voltaire**

After looking at enough brainteasers and optical illusions (see pages 00–00), by now you're pretty much convinced you can't trust your eyes. But actually it's your brain you can't trust, the interpreter of what your eyes take in.

Take for instance this simple drawing of a cube:

Interpreted in more than one way by your brain. Stare at this cube for a few minutes and watch it spontaneously "reverse" in depth. Concentration exercise: Practice holding one image as long as possible before allowing it to shift.

Figure 12.

If you stare at this cube for any period of time, you'll notice it seems to "shift," reversing its direction of depth. This isn't the fault of your eyes, it's due to the fact that your brain initially only "sees" (i.e., interprets) the cube one way, but eventually catches on and says, "Oh yeah, I can also look at the cube this way!"

This is known as "object fixedness," the tendency of the brain to see only one use for an object.

The brain always chooses the easiest and quickest path to identifying something. Only later, when the brain is sure the object poses no danger, does it relax and return for a second look. This explains why you jumped that time you thought you saw a snake in the grass but it turned out to be a piece of garden hose. Your brain "saw" the object, went straight to its "looks like a friggin' snake!" file and ordered your leg muscles to jump away from the (falsely perceived) "danger." Then, once out of danger, your brain took a closer, second, look and realized it's just a piece of harmless garden hose.

So sometimes we "see" things that aren't there. Other times our *untrained* brain misses things—important, possibly dangerous things—right in front of our face.

Another example, our eyes pass over a seemingly peaceful park setting and fail to notice the man standing in the shadows, leaning a little too casually against the side of a tree. When's the last time you got a "gut feeling" that something just wasn't right? Often this occurs because our eyes have taken in something novel or out of place that our brain pushes to a back burner, in favor of something the brain is currently concentrating on. You can see how dangerous this could be if our friend leaning against the tree over there has evil intentions.

So it's not like we're starting from square one in trying to improve our senses. We're actually starting from a deficit position! But with practice all things are possible.

Combat cadre and other professionals whose livelihood—and often their very lives!—depends on their "seeing right" the first time are trained that the human eye first sees movement, then silhouette, and finally color (Lung 2004b).

We too can train ourselves to "see" better and, more important, actually comprehend what we are seeing—or what we have been missing!

Sense of Sight Exercises

Review the illusions on pages 7–9 again.

Study every optical illusion book you can get your hands on—especially good, Al Seckel's *Incredible Visual Illusions* (Chartwell Books, 2005).

Learn how magic tricks work. Go to live magic shows. Expect to be tricked because you *are* going to be tricked. Try to catch the ol' switcheroo, the misdirection.

Consider this: At a live magic show you expect to be fooled and you're watching intently for the "trick" . . . and you still get taken. How much more so when you're *not* watching for it, not expecting it out on the street or in your place of business?

You need to take the same attitude with you from that magic show back out into the real world: You *are* going to be tricked. So watch for it. Train your eyes (and your brain) to see trouble coming. Paranoia can be a valuable asset.

Object-Spotting Exercise: Have a friend place a couple dozen small, everyday objects on the table between you. Now close your eyes while he adds or removes or otherwise rearranges these objects. Opening your eyes, try to determine what's different.

This exercise will help your eyes (brain) better spot when something is "out of place" or when a novel variable has been added to the landscape. Cheater's hint: With your eyes closed, listen carefully and you will be able to approximate where on the table your friend changes things.

Shape-Spotting Exercise: Walk down the street, or through a park during the day and try to spot all the objects shaped like, or composed of, circles. Now take the same walk and look for square shapes. Now try it for triangles—harder. How about diamond shapes? Harder still.

Retrace your route at different times of the day. Shadows look different in the morning than at high noon, than they do in the afternoon. What about at twilight, and again at night?

Notice how varying degrees of light and shadow—chiaroscuro, they call it—plays tricks on your eyes.

What you thought was a trash can turns out to be a squatting homeless man. Those two trees over there? One is that stranger you saw earlier in the day, but in the dark his silhouette resembles that of just another tree.

In the dark, new shapes—islands of light and circles of shadow—appear where none existed during the day. Shadows can hide a person standing just inside a doorway or trick the eyes into seeing false doorways—safe havens where none truly exist.

Your eyes are dazzled and confused by the changes and, as your mind struggles to "make sense" of contradictory information, your enemies close in . . .

Mind's Eye Exercise: Study this image for a full minute:

Figure 13.

Now close your eyes and practice keeping the "after-image" in your mind's eye for as long as possible. Don't be discouraged when the image eventually fades. That's normal. The goal is to keep the after-image in your mind's eye for longer periods of time.

In India, and points East, this kind of practice image is called a "yantra." Yantra range from the simplest—a dot, or circle—to incredibly complex images known as "mandala."

In practical use this exercise will train your eye to keep more complicated images (e.g., a park scene) in your mind's eye for longer periods of time, giving your mind a chance to file away its initial impressions and (re)examine the scene more closely for potential danger and/or opportunity.

"Restless man's mind is,
So strongly shaken
In the grip of the senses . . .
Truly I think
The wind is no wilder."
(Bhagavad-gita)

The Nose Knows

Our sense of smell is a too oft ignored sense. How often have we literally been led around by our nose, been manipulated by unseen forces that can influence our mood without our ever being aware of it?

Down through the years our olfactory sense has been targeted and twisted by military attacks, merchandising assaults, and out-and-out mind control ploys meant to manipulate our mood and master our mind.

And, in one way or another, all of these sucker-punches to our proboscis have been successful.

Military Use: Down through the years inventors have designed weapons and invaders have used weapons that specifically targeted our sense of smell. For example, ancient Chinese used "stink bombs" to unnerve their enemies (Seagrave, 1985:180).

Today the U.S. military, looking for nonlethal crowd control alternatives, have developed a noxious-smelling "puke gas" that, when dispersed over an unruly crowd, makes them vomit, urinate, and defecate on themselves.

Special Forces soldiers eat only indigenous foods prior to penetrating deeply into enemy-held territory where, even at night, their body odor (sweat, flatulence, feces, even their urine) can smell of foreign food, alerting the enemy to their presence.

In the modern war on terrorism, beyond bomb-sniffing dogs, we may now identify individuals (at airports, etc.) based on their distinctive body odor. Seems each of us exudes a combination of body chemicals as unique as a fingerprint (*Business Week*, August 8, 2005:55).

Marketing Use: Smells are being used to first draw us into stores and then "encourage" us to buy more while we're there.

Nike recently paid for a study that concluded that most people will buy more shoes, and be willing to pay a higher price for those shoes, if the room smells like flowers.

Likewise, the Las Vegas Hilton gambling casino found that its patrons spent 50 percent more time playing slot machines when the air around them was doused with a floral scent. The stronger the fragrance, the longer individuals gambled.

This even works in real estate, where the smell of fresh baked bread and cookies (giving potential buyers a "homey" feeling) increased sales (Ibid.).

To make matters worse, advertisers have discovered that by pairing the right packaging label to the right smell they can seduce the eye and the nose simultaneously.

A new study shows that identical scents smell different from one another, depending on the type of labeling used.

If a smell is paired with a pleasant-looking label, it will be perceived as a more pleasant smell overall than the same scent when paired with a more negative, less attractive label. It's not hard to figure out that our expectations influence how our brain perceives a smell. Research bears this out. Brain scans taken during this study reveal that the person's initial reaction—positive or negative—to the labels affected how blood flowed to the test subjects' olfactory processing areas.

Manipulating Your Mood: Above and beyond getting us to buy even more stuff we don't need, recent research into how we react to our sense of smell has led to several revelations, some of which might benefit us, all of which might be used against us.

- *Keeping us more alert:* The smell of peppermint or cinnamon can increase alertness, helping keep drivers awake. This according to a Wheeling Jesuit University study. (*Men's Health*, May 2006:36)
- *Improving our memory:* Smells that act as memory aids include rosemary oil, basil, lemon, and sandalwood—which increase both contemplation and creative thought.
- *Making you more successful:* People whose clothes smell of pine are perceived by others as being more successful, more intelligent, sociable, sanitary, and attractive than those whose clothes smell like lemon, onion, or smoke. (*Psychology Today*, September/October 2005:32)
- *Making you more sociable and trusting:* In research done at the Univer-

sity of Zurich, after test subjects sniffed oxytocin (a hormone associated with lactation and social bonding and interaction) they were 20 percent more likely to trust strangers with their money. (*Scientific American*, August 2005:26)

- *Improving your sex life:* Research has shown that between ages eight and sixteen, girls begin to dislike the odor of male sweat. Likewise, a recent survey of adult females found that women overwhelmingly agree that a man's body odor is more important than his appearance. (*Psychology Today*, September/October 2005:32)
- *Men also have noses:* A recent study rated unattractive women 20 percent more favorably when room was spritzed with pleasant fragrance. (Ibid.)

Some researchers believe that an inborn olfactory sense may act as a steering mechanism guiding men and women to members of the opposite sex (Bayer, 1987:34).

Recall from *Mind Control*, in the section on "The Art of Seduction," that when a woman is interested and/or aroused it may seem that the air between the two of you seems "thicker," as you subconsciously pick up on an increase in her sexual pheromones, the same way you "smell" rain coming.

Such human pheromones can now be bought off the shelf and are increasingly being added to perfumes and colognes. Also, many natural scents have been proven to affect us sexually. Sandalwood, for example, is considered an aphrodisiac by Hindu yogis.

FYI: Giving us the upper hand against our enemies: Prior to a meeting, douse the meeting room with scents designed to relax the person(s) you will be negotiating with. If that person happens to be a woman, depending on your agenda, you might also consider sexual pheromones.

Prior to your rival attending an important meeting, saturate his clothing with foul-smelling odors designed to be triggered by body heat.

Spray liberal amounts of a woman's perfume in his car, or onto his clothing, where his wife is sure to smell them.

It's no secret women have more sensitive noses than men (except for those guys on *Queer Eye for the Straight Guy*) so this ploy works even if you fold only a small amount of "the other woman's" perfume inside one of his handkerchiefs.

Sense of Smell Exercises

Practice smelling:. Fill your memory banks with different smells. Allow a friend to blindfold you and then try to identify different scents they place under your nose.

As with the taste exercise that follows, this can easily be used as a titillating, erotic game. (Who says sex can't be educational!)

The Z-E-N-Rose Exercise: On the other hand, the Z-E-N-Rose exercise is both a meditation and a way to increase your awareness of your sense of smell. When we are smelling something pleasant, a rose for instance, we draw our breath in to its fullest. Nostrils flared, we draw the air to the bottom of our lungs.

This is how you should breathe when you're meditating. And you should learn to meditate (1) to reduce stress, (2) in order to "center" yourself for more self-control, and (3) to increase your overall awareness, especially sense awareness.

Most meditation techniques call for you to "Sit in a comfortable, quiet place" . . . yeah, good luck finding a place like that these days!

This Z-E-N-Rose meditation you can do anywhere. A nice quiet place is great, if you have such a luxury. Filling your meditation spot with incense and other calming scents is also good. Real Roses? Even better.

Now take a minute to think about your breathing.

Close your eyes (if in an appropriate place to do so) and take in a full, deep breath, imagining that you are smelling a large rose, trying to draw as much of its delicious fragrance as deep as possible into your body.

As you breathe in, mentally repeat the letter "Z." In your mind, associate this "Z" with relaxing, with "getting your Z's," i.e., sleeping.

Hold this "Z" breath for a few seconds before exhaling. As you exhale, think "E" for "exhale." Without giving yourself a stroke, gently force all the breath from your lungs.

Now breathe in another "rose" breath as you mentally recite "N" (as in "breathing IN," get it?). Think about how an "N" is just another "Z" *relaxing* on its side.

As you breathe out this time, mentally say "ONE" . . .

Continue this breathing exercise, repeating "Z-E-N" as you breathe in, out, and in again. At the completion of each cycle of three breaths count "one," then "two," "three," and finally "four."

That's all there is to it. Probably take you three or four minutes, tops. Of course, you can do it longer if you'd like, if circumstances permit. Just start the exercise over once you complete "four."

And if you concentrate on your "rose breathing" before long, don't be surprised if you actually begin *smelling* a pleasant rose-like fragrance. This will be your proof that you have moved into a more relaxed state where the mind has created the smell from the image of the rose you have successfully held in your mind.

Calmer mind, stronger senses. Stronger senses, stronger mind.

FYI: A recent study at Massachusetts General Hospital concluded that forty minutes of meditation a day appears to thicken parts of the cerebral cortex. This is the part of the brain involved in attention and sensory processing (reported in *Psychology Today*, September/October 2006:74).

I Can't Believe What I'm Hearing!

The smallest bones in our body, the tiny bones in our ears, do one of the biggest jobs.

When the Native American put his ear to the rail, listening for the iron horse, he wasn't "hearing" so much as "feeling" vibrations. The same is true with the old woodsman's trick of sticking a knife into the ground to "hear" someone approaching.

"Hearing" is actually the bones in our ears vibrating in response to sound waves.

Old-fashioned safe-crackers didn't really "listen" for the sound of tumblers falling, they instead "felt" the tumblers clicking in place through their fingertips (hence the old movie cliché of a safe-cracker "sharpening" his fingertips on sandpaper).

Ever see a Hindu snake charmer, playing his flute, making the cobra dance? Well, first of all, since that snake doesn't have ears to actually hear the sound of the flute, the music is simply for the benefit of the audience. The snake is actually "charmed" by the gentle swaying of the fakir and the tapping of his foot—which the snake "hears" vibrating through the ground.

Human capacity to hear changes over time, and not just in old age. For example, the cochlea in the ear begins to "dull" after age twenty-five. As a result, "kids" can hear a least one frequency we know of that older people can't. A British department store isolated this frequency and used it to pipe "uncomfortable" music aimed to discourage kids from loitering (*Good Morning America*. ABC News 6/13/06).

As with the eyes, our ears pick up a lot more sounds than we are ever consciously aware of, multitude of sounds, most of which we ignore, as our brain sorts through and prioritizes what it considers immediately important. This explains the "cocktail party effect": how in a crowded, noisy room, we still immediately notice when someone casually speaks our name from halfway across the room. That's because our name is important to us and our brain take immediate notice of it being spoken.

Black Science smoozing tip 464: When meeting people, especially for the first time, always smile and repeat their name, perhaps even adding a flattering comment about their name.

People like hearing their own name. So it stands to reason they will like someone who says their name often, while smiling, always in a flattering, interested-in-you-as-a-person way.

Sense of Hearing Exercise

Listening to Music: Pick one instrument and concentrate on it. Try to isolate the drum, for instance. Then do it again for the guitar, then the keyboard. This is easy to do with loud rock music, harder with more subtle arrangements. Orchestral music is excellent for this exercise.

Practice Isolating Sounds: We can more easily "hone in" on conversations at the next table over. Learn to lip-read.

The Farthest-Away-Sound Exercise: Begin by first using our Z-E-N-Rose exercise to relax yourself. Having accomplished this, now close your eyes and really listen to the sounds around you.

After a few minutes, gently "push" your thoughts further outwards, as if "a circle of sound" is expanding out from you on all sides. Listen for the farther away sounds—sounds outside the room you're in, perhaps outside your house.

When doing this exercise outside, on your porch or in the park, listen first for birds in the area, then those farther away, and then farthest away. Perhaps you'll hear that squirrel over there playing in the fall leaves, or else scratching his way up a tree. Extending your listening farther . . . now you hear that jet high up in the sky, and a subtle sound of the wind picking up.

Of course, you should also be able to *feel* this breeze on your skin, perhaps *smell* the rain coming.

Leaving a Bad Taste in Their Mouths

You wouldn't expect Imperial Chinese poison tasters to have had much of a retirement plan. But the truth of the matter is many of them were experts in their craft, some able to discern with just the minutest taste whether something was safe for the Emperor to partake of.

Those tasters not so accomplished didn't live long enough to pass on their skill.*

Sense of Taste Exercise

Improve Your Sense of Taste: Do this exercise, similar to the exercise we did to improve our sense of smell. Blindfold yourself, and have a friend give you small amounts of food—and other nontoxic materials—to taste. Note the texture of these foods, not just the taste.

FYI: With the right person, this exercise makes great foreplay!

A Little of That Human Touch

Your skin is your largest organ. (Insert the ribald joke of your choice here.)

Human skin is so sensitive we can sometimes feel the weather changing (air pressure, increase of moisture in the air, etc.). We can feel a potential lover's body heat increase along with their arousal. In return, we show our interest through our sense of touch: by smiling and lightly touching her arm—or her, your leg. When someone touches us in this way, we feel flattered, comfortable with them, and feel like we are more attractive. Recall from *Mind Control,* "The Art of Seduction" how the quickest way to a woman's heart is through her elbow—touching her elbow, that is.

We can spy out someone's passing or their presence by sweat left on a doorknob, by heat left on a chair, or by feeling the hood of a car to tell if the engine has recently been running.

---

*It also helped that some of these poison tasters took small amounts of known poisons over the years, slowly immunizing themselves against commonly used poisons.

Chinese masters of traditional Chinese medicine can often diagnose a person's overall health simply by feeling the patient's pulse and/or by using their hands to determine the person's *chi*-flow. And, yes, such knowledge was later used to develop the martial art of *Dim Mak*—"death touch."

In much the same way, a Western doctor might feel a patient for blood flow to an extremity (coolness and rigidity in the limb an indication of poor circulation).

Chi masters can not only diagnose disease, they can also ferret out psychological stress by touch—recognizing tension and/or spasms, tics and twitches, in muscles.

In the same way, adept Western chiropractors are also sometimes able to intuit psychological tension as the basis of physical maladjustment.

It's not as hard as it sounds. I'll bet you can already tell when someone is worried or scared just from their sweaty palm and/or from the slight tremble you feel when shaking their hand?

Some secret societies, Freemasons for example, have special handshake grasps that allow them to feel the pulse of the other person, in order to tell if the other person is unduly tense and even lying (see figure 14).

Modern-day magicians, those who specialize in "mind reading," are actually only adept at reading the tension in a person's shoulder, arm, or hand while that person is being asked to think about under which of several cups they've hidden an object, which the magician then finds.

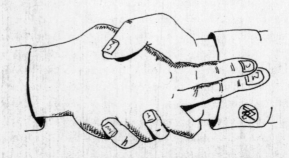

Figure 14. Masonic "Pulse-reader" Handshake.

As the handshake is formed by the little finger of each hand slipping between the little finger and ring finger of the opposite hand, the index finger and middle finger slip inside the cuff to feel the other person's pulse.

Accomplished ballroom dancers use their body-sense of touch to feel their partner's every move. Likewise, martial artists, especially those who specialize in close-in fighting (e.g., *Judo* players and *Aikido*), must instantly react to an opponent's shifting position, telling by touch—feeling the subtle shift of balance, a settling of their opponent's weight, etc.

During the Middle Ages, *Shinobi* Ninja, anticipating having to fight enemies in pitch dark, would strip themselves naked. That way, whenever they touched clothing, they instantly knew it was an enemy.

Sense of Touch Exercises

Tai Chi and Chi Gong: Learn and regularly practice tai chi and chi gong. When dancing your hands through the air during these exercises, imagine you are pushing your hands through warm water. Try to feel the still air on your skin.

The Go-Ju *Exercise:* (Japanese for "hard and soft") is simply carrying a small hard object, a rock for instance, in your right pocket and a soft object, a small piece of fur or wool, or one of those rubber-squiggly balls (the ones with all the tentacles on them) in your left pocket.

Now when you feel stressed out, slip your hand in your left-hand pocket and squeeze the soft object several times.

Conversely, when feeling like you need a short burst of energy, or when you feel the need to "center" yourself, slip your hand into your right pocket and grind the rock or other hard object into the palm of your hand.

Too simple an exercise, you think? Reconsider.

The mind operates on very simple signals. Merely rubbing your hand across something soft (1) makes your mind identify and record the experience, and (2) in the process of identifying and categorizing the object in your hand, your brain sifts through other objects in its files labeled "hard" or "soft," thus just one squeeze of such an object causing a dozen images of "hard" or "soft" to race across the mind, helping establish, or reestablish, a calmer or else energized and determined mind-set, respectively.

Making Sense of the Sixth Sense

This is where all the trouble you've gone to to increase your awareness of, and acumen with, your five senses begins to pay off.

Having increased our awareness of the five senses, it will often seem to others, and sometimes even to ourselves, as if we have access to a sixth sense—ESP—simply because we have become more attuned to picking up on unconscious clues and cues given off by others. We see divulgences and discrepancies from their body language, we hear hesitations in their speech, feeling their body tremble or noting their sweaty palms when we shake their hand. Perhaps even smelling the lust—or fear!—exuding from their pores.

Our five senses now serve us well. And often these five "sense awarenesses" are the only edge we'll need to confuse, defuse, and ultimately defeat an enemy.

But what if we could take these five sense awarenesses to the next, deeper, level? Wouldn't it be great if we could actually read an enemy's mind, know what he intended before he even had a chance to set his plots and plans into motion?

A July 2006 *USA TODAY* survey asked adults what "superpower" they would most like to have: 1 percent said they'd like to be able to walk through walls; 11 percent wanted to be able to turn invisible; another 15 percent dreamed of having the ability to fly.

But the number one "superpower" most people chose, 28 percent of those surveyed, was the power to read minds.

According to some recent scientific findings, this may not be such a farfetched wish.

We're not talking about ESP, some "psychic" extrasensory perception. This is *ASP—additional sensory perception.*

At the basic level, ASP begins with the full use of our five senses, as already mentioned, to the point to where it's giving the impression that we possess extrasensory perception.

But ASP then goes on to incorporate recent discoveries about how our five senses gather information, and more importantly how we process that information to the fullest extent of our five senses—and beyond!

For example, some people who have lost their sight can still "see," possessing the ability to "guess" the movement, color, and shape of objects around them, getting the answers right most of the time. This "spooky sense," as it's sometimes called, is more commonly known as "blindsight."

Blindsight occurs when the eye takes in images that somehow bypass the (damaged or cut-off) primary visual cortex (the part of the brain that sifts through data from our sense of sight). And even though we are not consciously aware we are "seeing" such images, other parts of the brain may process (respond to) this "shadow" information without waiting for it to be processed through normal channels, i.e., through the primary visual cortex.

One theory is that "blindsight" is a good thing, preventing the brain from going into "sensory overload" by "filtering" the vast amounts of information coming into the brain via the eyes at any given moment. Thus, at any given moment, our brains are receiving—and responding to—not only a flow of conscious "sensory" information, but an "extrasensory" flow of data as well that we are nor aware of. As a result (Kruglinsky, 2006):

> The unconscious flow of information . . . allows us to change our behavior and make decisions without ever quite knowing why we did. (p. 13)

Thus we find ourselves unconsciously making decisions, adjusting our behavior to fit information that we are not consciously aware we are receiving. Furthermore, we often think we "predicted" something was going to happen—perhaps avoiding disaster—even to the point of congratulating ourselves for being "psychic" when, in reality, it wasn't ESP, it was ASP—the additional uses of our already existing senses' ability.

Seeing us act or react to unseen data, others might also be in awe, also believing us to have ESP.

I'm sure you can see the advantage in allowing our enemies to go right on believing something like that?

Another recently discovered natural process that might give the impression we possess ESP involves what are known as "mirror neurons".

According to author Ker Than:

> Buried deep inside your skull are special brain cells that read the minds of others and know their intentions. (*Psychology Today*, August 2005:26)

These mirror neurons are brain cells that fire in response to the "reflection" of another person. For example, when you watch a coworker lift his cup to take a drink, the mirror neurons in your brain activate just like they would if it was you lifting your cup to take a drink.

Neuroscientists believe these mirror neurons are what make us feel empathy and compassion for other human beings and cite research showing that autistic boys' mirror neurons fail to fire in this way, and may account for an autistic child's lack of social interaction and communication skills (Than, p.26).

Recall that some Hindus equate this sixth sense with the use of the heart, i.e., feeling and compassion, which we might also think of as simply being empathic, which brings us right back to those recently discovered "mirror neurons."

As science continues to uncover such new information, or at the very least continues to find credence in wisdom intuited by mind-masters of long ago, we may one day find that all of us do indeed possess some sort of ESP—"extra senses potential" beyond the five we most often use.

But until then, we need to put effort into developing the first five, if only to show we deserve a sixth.

Six Sense Exercises

Repeat the exercises for the eyes, the ears, the nose, your sense of touch and taste. And then repeat them again.

Sensory Modes

One final way in which we can use our knowledge of the five (or is that six?) senses is to learn to discern which of the five senses others are dominated by.

While we all use all our senses to varying degrees, we still each tend to favor one sense over the others.

This unconscious preference is often reflected in our choice of entertainment—going to a movie (seeing) versus going to a music concert (hearing), versus going to a dance (touching). Our preference is also reflected in our hobbies and in our choice of careers.

A taste-oriented (-dominated) person might be right at home as a chef. Ah, but they could also like working with their hands (touch oriented) which is also used a lot in cooking.

Though we usually associate teachers with talking, those dominated by the sense of hearing actually make the best teachers since they can better determine a student's individual needs by really listening.

We associate artists with the sense of sight, but what about all those sculptors and carvers dominated by the sense of touch?

While being mindful not to too quickly classify (i.e., "lock in" or stereo-

SIGHT
I see what you mean.
Looking out for #1.
Apple of my eye . . .
Keep your eye on the ball
Try to see it my way
Get the Picture?
See where I'm coming from?

SMELL
Something's rotten in the state
of Denmark . . .

Something's fishy . . .
Something doesn't smell right.
Stinks to high heaven!
I smell a rat!
Shitty deal . . .
Garbage!
Bullshit/BS.
Sniff out an opportunity.
Need breathing space . . .

HEARING
I hear you loud and clear.
That's music to my ears.
I like the sound of that offer.
You're talkin' my language!
I'm all ears.
Get an earful of this . . .
Pump up the volume . . .

TOUCH
It just feels right to me.
Turn up the heat, make 'em sweat!
Heavy costs.
Sticky situation/firm grasp
Too hot to handle/hot potato
Smooth operator
Build on our relationship . . .
Solid idea/carry the weight
Let's shake on it . . .
Close-knit family
Get wind of . . . cool deal

TASTE
Leaves a bad taste in my mouth.
Sweet deal . . .
Sour grapes.
On the tip of my tongue . . .
Let 'em chew on that a while.
Nibble.
We can lick 'em!
Get a bite of this offer . . .

Figure 15.

type) a person into one of these five sensory modes, it is always good to remember that people sometimes use more than one sensory mode of talking, just as they use different senses from situation to situation.

Most often we can discern a person's dominating sense mode simply by the words and phrases they choose to express themselves with. (See figure 15.)

Once we figure out which of the five sense modes dominates our target person, we can then craft our speech, etc., to fit him, in order to better smooze, i.e., kiss ass.*

You can also use knowledge of someone's dominant sensory mode to toss a stumbling block in front of a rival.

For someone who is "seeing" oriented, arrange for them to have to listen to lengthy, boring audiotapes.

Once you peg his sensory mode type, it is a simple matter to craft a counter strategy.

---

*Black Science adage number 169: It's not "Kissin' ass" if you know you're kissin' ass. If you know you're kissin' ass . . . then it's strategy!

Ping-Fa Strategy

"No destiny but that which we whittle with our own will, craft with our own cunning."
—Duke Falthor Metalstorm

Having first cleansed the body of weakness in order to make it strong, having then awakened the senses through *Jing Gong*, Red Spear students (those who survived this far!) now advanced to the cerebral "Hall" of learning—the *Ping-Fa* strategy level of training.

When we speak of Sun Tzu's *"Ping-Fa,"* we are referring specifically to a treatise, an "art of war" opus, a master's text. But *Ping-Fa* can also refer to the overall art of war—the craft and the cunning, mind-set, machinations and maneuverings it takes to survive and thrive beyond your enemies.

Red Spears had an eight-point system for teaching strategy, a comprehensive system that had its inspiration in the Pa-Kua "Eight-Trigrams" practice that was first a philosophy that later martial art style developed by Taoist alchemists at the Yu-hau shan monastery.

The martial art of Pa-Kua uses open hand strikes paired with circling-around footwork that forces an opponent to constantly shift his position, denying him a firm stance from which to launch counterstrikes. In other words, you keep your enemy off balance at all times.

The Red Spears' Ping-Fa strategy is built on this same principle, only it's your enemy's mind you keep off balance.

This strategy also values the astute observation, realistic evaluation, and succinct application (and, if need be, reassessment) of a plan. Thus, in many ways it can be considered a more intricate model of the typical Western "problem solving" model: (1) clearly define the problem, (2) brainstorm possible solutions, (3) prioritize possible solutions in order of their best chance of success, (4) implement prioritized solutions, and (5) adjust as need be to compensate for shifting realities.

The eight steps of Ping-Fa strategy, illustrated in figure 16, follow.

1. *Chou* (Chinese "to measure"). Before embarking on any campaign, mission, or business venture for that matter, we must first consider—"measure"—two variables: *Xing* and *Shih*. Xing and Shih (pronounced *shing* and *she*) are another way of saying circumstance and flux.

Xing is the outward appearance (shape) of a thing, a person, or a situation—the way it "appears" to be.

In the case of a person, Xing is the proper, public face they show to the world.

Shih, on the other hand, deals with the essence of a thing, the potential or stored energy within. In the case of a person, Shih is their real face, the one they keep hidden from the world. Expose this face and you expose your enemy for what he truly is.

Shih is the inner dynamic, the inherent power of something. This includes all the possible outcomes to a given situation.

When "measuring" people, Shih reminds us of Sun Tzu's "Know your enemy." But since you are a vital counterbalance to any such measurement of your enemies, Shih also reminds of the second part of Sun Tzu's teaching, ". . . and know yourself."

In any given situation, we must first "measure" the potential of ourselves. Do our resources and determination—i.e, ruthlessness!—coincide with reality?

Based upon his past actions, what is our enemy capable of? And what is the current situation, i.e., is this the right time and place to attempt this?

All these must realistically be weighed—measured—before we continue crafting our strategy.

2. Suan ("to calculate"). Having taken stock of the times, situation, and assembled dramatis personae, we now "calculate" Li versus Hai.

Li is "advantage," all that we have to gain if we prove successful in our venture, if our strategy and tactics bring our enemy to ground, if we should succeed in snatching the brass ring.

Hai, on the other hand, is all that we are willing to risk, all things precious we have just placed into the uncertain hands of fate. This is the cost we are willing to pay to carry the day, to be able to stand with our foot firmly on the crushed chest of our foe while baying at the moon—or, at the very least, while ravishing his woman and raping his bank account!

3. Ji ("planning"). Having measured our worth and the worthlessness of our enemy, having then calculated what we are willing to risk in order to gain a thousand-fold, we now come to the planning stage.

It is at this point we can begin setting up our little toy soldiers and tanks on the practice map, mirroring our actually beginning to surreptitiously move our men and materiel into place. It is at this point we start drawing our hopeful lines and arrows on the map—the ones that lead from our staging ground straight down our enemy's throat!

It's all about logistics: How to get your soldiers from point A to point B with full stomachs and plenty of ammo; how to get your idea off the drawing board and into the bank.

When possible—and when prudent?—we consult with experts in the field; veterans who know the landscape we're venturing onto, be that backdrop financial or homicidal.

4. *Ce* ("to scheme"). Having laid out our plan—our routes of approach, how much toilet paper our legions will require during the course of the campaign—we now need to think about what our enemy is thinking.

Whereas "planning" focuses on those "necessities" in life, those things we have to have, the basics (a full tank of gas and a road map, investment capital, etc.) before starting out, "scheming," on the other hand, is where we start "tweaking" that plan, turning an average attack strategy into a Pearl Harbor—or a Nine-Eleven.

The enemy has spies, too. He sees you pooling your resources, training your troops, amassing supplies—he knows you're up to something and will soon be busy making his own counterplans.

Ce—scheming—also involves the conciliatory gestures and camouflage you'll use to reassure him nothing's really going on, that you're not up to something. Ce can also involve distracting him with your left hand while you slap him with your right!

I'm sure you've heard about all those full-size *wooden* planes, tanks, and other equipment—complete phony airports and fake army bases replete with straw soldiers—that were set up along the English coast designed to confuse Hitler's spies just prior to the D-day invasion? Well that's a good example of Ce-scheming.

5. *Zheng and Qi* ("direct and indirect" aka *Cheng and Chi*). At this stage we are forced to take a long, hard look at both sides of the coin,

forcing ourselves to listen to counterargument and naysaying we don't want to hear—all the possible things that could go wrong:

"Nay-sayers seem capable of saying naught but Nay, yet if they could say anything else in any other way they'd eventually make that too a Nay. So, listen not, for Nay is but all the Nay-sayers would say." (Andre Spearman)

Granted, it would be nice if we were already the genius strategists who could foresee every possible contingency and map out our attack accordingly—without help.

Not gonna happen. Some of the greatest military minds in history have both won and lost great battles based purely on luck.

Of course, if like Dr. Lung you don't believe in *luck* . . . chalk it up to "shit happens!"—either way, some things will forever remain incalculable and unforeseen. That's why we learn to think on our feet, roll with the punches, and always keep a spare pair of underwear handy.

And, yes, we never get tired of repeating Clausewitz's mantra: No battle plan survives first contact with the enemy.

We adjust . . . or we just die.

No man's good enough to be his own lawyer, that's why every good leader surrounds himself with a crew, a posse, a hardy bunch of henchmen— brothers of the spear not afraid to express their doubts, not afraid to tell the boss when they think he's messing up.

During his early campaigns, Alexander the Great kept a loyal cadre of companions, mostly boyhood friends who had grown up fighting beside him. Together, they conquered a goodly piece of the world. It was only when Alexander no longer had those comrades to give him advice that he began to weaken—when he had no one left to tell him when he was getting ahead of himself, getting too big for his britches (or whatever those little skirts they wear in Greece are called).

Likewise with Adolf Hitler who, in his early days, had practical "old school" soldiers like Ernst Rohm (head of the SA—Stormtroopers) to remind little Adolf where he'd come from. But as Hitler assumed more power by currying the favor of mercenary financiers and self-serving politicos, Hitler betrayed and executed Rohm and others who'd literally fought their way up from the streets with him, replacing them with kiss-ass yesmen, spineless sycophants either too afraid to question Der Fuhrer or else too caught up in their own power-mad fantasies to point out to him those sure looked like American tanks barreling down the autobahn!

In today's parlance: You need a good posse to watch your back, to tell you when you're screwing up.

Zheng and Qi also apply to how we deploy our forces and to how we actually fight a battle.

There are direct (Zheng) forces we can use, and direct, straightforward challenges we can aim at our rival; meeting him man to man, face to face on the field of battle. But that's not always practical even if we do know ourselves to be evenly matched to our opponent.

Never go toe to toe with an opponent unless you *know* you are in the superior position—and, even then, try to avoid such bravado.

Qi means "indirect" and refers to guerrilla tactics we adopt when we know our enemy has superior numbers, superior firepower, or holds a superior defensive position—one that would tax our resources to dislodge him from.

In such an event, we play the gorilla, nipping at his heels. Unseen we quietly undermine his tower wall. We slowly, secretly, unsuspected, gather together our forces, increasing our worth and power incrementally, so as not to arouse suspicion; building up our reserves, biding our time until the time is right to strike.

When one big business plots the "hostile takeover" of another company, before revealing their intent, they quietly start buying up shares of the targeted company—buying these shares secretly through third parties, so as not to alert their intended victim.

This is no different from the kinds of secret training and slow, subtle shifting of resources Hitler accomplished prior to launching World War II; successfully rebuilding Germany's military might right under the noses of those who should have been watching him the closest.

Of course, you don't have to choose between using one or the other, Zhing or Qi. In any conflict or campaign, a good general uses both.

In his *Art of War* Sun Tzu teaches not only how a general moves his conventional (Zheng) troops from place to place, he also stresses the importance of a wise general employing plenty of unconventional (Qi) forces: fast-moving shock troops and spies.

A good sword cuts both ways.

6. *Jie* ("timing"). The best product in the world won't sell if the people aren't ready for it.

Hitler invaded Poland in the fall, gambling that he could be in Moscow before the same winter snows that had smothered Napoleon stole the thunder from the German blitzkrieg. Bad timing, Adolf.

Jie also means "space," reminding us to take note of the distances between things, people, and destinations.

In martial arts, students must master a concept called "bridging the gap," which means closing the distance between you and your enemy as quickly as possible and taking him out as efficiently as possible—preferably with one blow.

This same "bridging the gap/take your enemy out with one blow" attitude is what you need to take with you into all your dealings. In other words, cross questionable and dangerous ground ASAP to reach your target-goal on the other side.

Of course, *Jie* also cautions that there is some ground not to be crossed, some hills not to be contested, some fortresses too costly to besiege.

1863. Gettysburg. Pickett's charge across a mile of open ground. The beginning of the end for the Confederacy.

7. *Gancui* ("to penetrate", pronounced gang-kway), literally translated means "to penetrate neatly and completely". This refers to getting inside your enemy's head in order to (1) know what he's thinking, and (2) Tell him what he should be thinking, to use deception to sow daunting challenges and doubt into his mind.

Gancui also requires penetrating *deeply* and *completely*. When we strike, we give it all we got. No half-assin'. In karate they call this *Ikken Hisatsu,* "to kill with one blow."

And while we may not be using a physical Tsuki-punch or Shuto-swordhand strike to bring down our opponent, when striking with our "Mind-Fist" the same rules apply.

We first steel our own mind before we steal our enemy's mind.

There is a story in the *Gospel of Thomas* that tells how a man, determined to kill his enemy, first practiced thrusting his sword into the wall of his home until he knew his arm (and mind) were firm enough to do the deed.

This should be our *gancui* attitude in all undertakings.

8. *Sheng and Bai* ("victory & defeat"). In the end we win or we lose. But it's not always that simple. There's a thing called a "Pyrrhic victory," a win that costs you too much to win. (It's like saying "The operation was a success . . . but the patient died!", or like staying in a toxic relationship just because the sex is good . . . or at least available.)

A victory that depletes your resources leaves both you and your defeated foe vulnerable to attack by a third party who has been waiting patiently for you and your enemy to cut each other down to a size he can swallow!

Of course, this is a great opportunity, provided you are that patient third party.

So even after you win your victory, that's no time to drop your guard with premature celebration.

During any military campaign setbacks and stumbling blocks magically appear that couldn't possibly have been anticipated during the planning and scheming stages. Likewise, sudden advantages can also appear out of nowhere that if not grasped immediately evaporate just as quickly as they appeared. Opportunity lost.

After any "mission" comes the "debriefing," that after-game chat with the coach when you take note of what went right, and what you could do better next time; about how the opposing player breaking his ankle was a "godsend" when your team was down by 20 . . . and how you can't depend on "godsends" to save your ass every time.

If you'd lost you'd damn sure be trying to figure out, "What the hell went wrong!" Do the same when you win; sharpening your game for the next time, insuring you'll win by an even greater point-spread next time . . . improving your chances of walking away with the championship even in the event "Lady Luck" gets distracted by the peanut vendor.

And, even when you win victory, instead of resting on your laurels, immediately turn your mind to measuring *Chou* again, honing your talents, stockpiling your weapons, readying yourself for the next battle, sniffing the air for fresh game.

Mother Nature is a fickle and insatiable lover, it takes a lot to keep her amused.

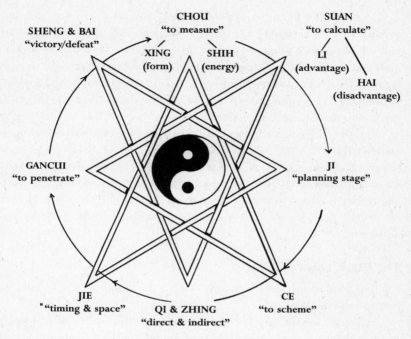

SHENG & BAI
"victory/defeat"

CHOU
"to measure"

SUAN
"to calculate"

XING
(form)

SHIH
(energy)

LI
(advantage)

HAI
(disadvantage)

GANCUI
"to penetrate"

JI
"planning stage"

JIE
"timing & space"

QI & ZHING
"direct & indirect"

CE
"to scheme"

Figure 16.

Peace is closer to a coffee break than it is to recess. There truly is no rest for the wicked . . . nor do we want any!

THE DEADLIER OF THE SPECIES

"Even the most beautiful woman in the the world, if she stands beside a lake will frighten the fish, and if she walks through the woods the deer will be startled and flee."
—Ancient Chinese saying

It has oft been argued (mostly by women) that "history" is just that: *his-story*, that the womanly gender has been woefully under represented in history. While this may be true overall, when it comes to plying the Black Science, female adepts are not hard to come by—both East and West.

The Bible, for instance, is full of scheming wives, witches, wenches, and women warriors beginning with Eve ("I was framed!"), down through that horny Egyptian wife who tried to jump Joseph's bones; from dauntless Deborah and dazzling Esther, to that "bad to the jawbone of an ass" barber Delilah. There's king-wooing Bathsheba, head-hunting Salome, and Jezebel, whose name has become synonymous with the kind of woman you definitely don't want to take home to mother!

Outside religious mythology, we find Helen of Troy—near whom men couldn't think straight; Lucrezia Borgia—who couldn't keep her poisons straight; Elizabeth, the "Virgin Queen"—who still somehow managed to keep Sir Francis Drake (and Shakespeare?) straight; horse lover Catherine the Great, and Hillary Clinton—who rode a jackass all the way into the White House!

In the East as wall there were—are—women both noteworthy and notorious.

The Black Lotus

The ideal of the Eastern beauty—the geisha, the concubine—will forever *fascinate* the West.

While we cannot help but be captivated by their beauty, their demure manner, there are all too many historical glimpses of the will and wile—and perhaps wickedness—lying just below the surface, behind the veil, beneath the silk?

In fiction, there's the melodic-voiced bride Scheherazade of *1001 Arabian Nights*, mesmerizing her husband with her tales.

On of the most dreadful killer cults ever spawned, the dreaded Thuggee of India, worshipped not a manly war-god, but rather the black-tongued, fierce-eyed bitch goddess *Kali*. (See Lung, 1995.)

Then there's the insightful twelfth-century Tantric Master Mahadeviyakka we talked about in chapter 6.

In the twentieth century, Indira Gandhi ruled India willfully well, until her assassination in 1984.

The Chinese had their classical heroine Mulan, as well as scores of twentieth-century Red Guard women warriors.

And while we always hear about the "Brothers" of Shaolin, that Order had a nun's branch as well.

In fact it was a Shaolin Buddhist nun who taught runaway bride *Yim*

Wing Chun Shaolin boxing, allowing her to fight her way out of a forced marriage. Today there is a martial art that still bears her name: Wing Chun, the art that launched Bruce Lee's career.

Like Wing Chun the martial art, women in China (as perhaps women everywhere) mastered the concept of *Shun*, which literally means "compliance" but which implies "going with the flow."

Knowing they couldn't fight a superior force, superior strength, they adopted the philosophy of "give way in order to get your way," the Judo principle.

Chinese women didn't need Sun Tzu to tell them: "When strong, appear weak."

Nonetheless, Shaolin nuns also took Sun Tzu's philosophy, the fighting arts they had learned inside the monastery walls, and everything else they could carry as they fled from the 1644 Manchu invasion.

Like their brother monks, those nuns who survived the Manchu massacre of the Shaolin Order are credited with (or else stand accused of) having founded their own triad—the Black Lotus—secret society (Seagrave, 1985):

> There had long been a female secret society in China, with branches throughout the rest of Asia, organized for the exclusive purpose of assassinating or otherwise punishing men. (p. 261)

That various triads had female members, or at least female auxiliaries, and that they often used female agents is well known. That a separate triad—perhaps several loosely linked female triads—existed, and are rumored to still exist in China and in other parts of Asia, is not that widely known or admitted.

Most trace the existence of this secretive sisterhood to the fall of Shaolin, but others believe the Black Lotus—but one of many names it has been called, e.g., "Celestial Sisters," the "Ever-empty Cup," the "Silk Touch"—had its true origin far before the founding of, let alone the fall of, Shaolin.

These researchers find hints, allegations, and veiled references to the existence of such a sisterhood stretching far back into China's history, with influences flowing both to and from all other parts of Asia.

Those researchers with a decidedly more "occult" bent have gone so far as to accuse the sisterhood of supernatural origins, something the Black Lotus did nothing to discourage.

In the same way China's *Moshuh Nanren* "ninja" never discouraged the belief they were descended from *Lin Kuei* "forest demons," so, too, Black Lotus sisters never discouraged tales tracing them back to, or at the very least placing them in league with, the much feared clans of *Caibu* . . . vampires!

Pronounced *T'zee-boo,* sometimes written T'zi-bu, Caibu aren't bloodsuckers like Western vampires, they're "psychic vampires" who live off the life force of others. Indeed, the name means "taking the essence of another."*

Caibu are more comparable to the Western belief in succubus—demons who seduce men by appearing in the guise of beautiful women.†

That beliefs linking the Black Lotus to supernatural goings-on would be perpetuated (outside of sisterhood propaganda, that is) testifies to the trepidation with which common folk, and the authorities, came to view even the mere mention, let alone the actual existence of, such a secretive order.

That this sisterhood existed prior to Shaolin is likely, given China's proclivity toward secret societies in general. But as to whether the Black Lotus originated in China or was imported from elsewhere may forever remain obscured behind a silken black curtain of charm, contrived coquettishness, and craft.

On a more positive note, we do know that similarly secretive sisterhoods existed in Japan for centuries.

Shinobi Ninja clans trained and fielded *kuniochi*, female agents every bit as deadly as their male counterparts.

Yakuza also employed women—agents, prostitutes, perhaps your cleaning lady!—to gather information, to leave the right door unlocked at the right time, to slip a little something extra in your morning tea.

"The Floating World" of the Geisha was also an exclusive sorority—keeping their own confidences, purring pleasantries, while lapping up the confessions of their clients.

And then there were the Miko.

The Miko

The Chinese phrase *Shen-tao*, meaning "spirit way," becomes *Shinto* when written in the same characters in Japanese.

---

*See *Mind Control*, 2006, page 244, for a more detailed discussion of Caibu and their connection to the Moshuh Nanren.

†Lest we discriminate, there are Iccubus who perform pretty much the same services for women.

Shinto (lit. "way of the gods") is the ancestor worshipping state religion of Japan whose roots can be traced back to ancient animism and shamanism.

Shen-tao originally referred to spirits and to spirit worship. But, like so many things Chinese, Shen-tao was eventually adopted and adapted to fit Japanese sensibilities.

That Japan owes much of its early development to China is well documented. Japan's bureaucracy, much of its religious thought—particularly Buddhism, many of its arts and crafts, came from China.

Political intrigue, already a Chinese specialty, also found its way to Japan. Some say, had its way with Japan.

For example, prior to the sixth century, the ruler of the Yamato clan (they weren't called *Shogun* yet) was only a local warlord, wielding power only over a small, contentious league of clans. But during the sixth century, reaping the bounty of seeds secretly sown centuries before, the Yamato made their move, and—eventually—succeeded in either destroying or otherwise dominating all the other clans, uniting them into a single kingdom. Think King Arthur.

This kingdom, united under the iron hand of the Yamato, eventually became the Imperial clan, the clan from which the future emperors of Japan would spring.

Before long the Yamato king had adopted a new title, taken from the Chinese: Ten'o—"Heavenly Emperor"!

But most experts agree that the foundations for this Yamato unification had been laid three centuries earlier.

In the third century CE the crown of the Yamato leader was seized by a woman. Her name was *Himiko* (sometimes written *Pimiko,* or simply *Miko*).

Himiko was a powerful shaman, believed by friend and foe alike to possess magical powers. Somewhere between casting spells with her left hand and wielding a sword in her right, Himiko rose to be undisputed leader of the Yamato.

But not only did she become ruler of the Yamato, mater familias of the Imperial lineage, she also became the spiritual founder and inspiration for a secretive shamanistic sisterhood whose name derives from her name: Himiko-Miko!

Miko are female shamans and psychics who possess knowledge of *Kogaku,* "the ancient learning," believed to be a form of powerful nature magic.

Some Miko are *Itako* (aka *Ichiko*), women shamans (often blind) and spirit mediums who act as human oracles (*Takusen*) through which *Kami* (nature spirits) and the spirits of the dead instruct the living.

Like their Greek counterparts, it's not hard to see how even a minor Miko oracle might all too easily influence events—alliances, state marriages, even whether or not a clan would go to war.

How much more so a strategically placed Miko like Himiko?

In fact, some suspect the meteoric rise of the Yamato to have been a plot by Black Lotus backers to take advantage of Japanese belief in the power of their shamanistic Miko to set up a counterbalance to the Chinese Imperial Court—allowing them, in effect, to rule Japan by proxy. Others go so far as accusing Himiko herself of having been a member of the Black Lotus.

Similar allegations of having been a Chinese agent provocateur have been lodged against Japan's first ninja *Otomo Shinobi,* spy war confidant to Prince Shotoku during his sixth-century war of succession.

Miko still exist today in parts of Japan (Bocking, 1997):

> It is a local custom that all women in Okinawa who have reached the aqe of 30 are initiated as NANCHU (the equivalent of MIKO) in a solemn ceremony called IZAIHO held every twelve years (November 15–18 by the old lunar calendar). (p. 146)

Whether the Miko of Japan, and similar sisterhoods scattered throughout Asia—some serene, some sinister—were birthed by the scheming of the Black Lotus depends on which scroll you read.

Some say it's vice versa: with elements of Miko and other feminine Asian fellowships influencing, if not the origin, then at least the subsequent development and diffusion of, the Black Lotus's craft and confusion.

Whatever the truth, wherever the locale that that truth is being touted or tested, one thing remains constant, and that is the overall art of strategy and tactics employed by such sisterhoods.

Theirs have always been, and always will be, first and foremost, truly the art of seduction.

The Art of Seduction (Part II)

"God hath given you one face and you make for yourselves another:
you jig, you amble, and you lisp, and nickname God's creatures,
and you make your wantoness your ignorance. Go to, I'll not more
on't; it hath made me mad."
—Hamlet

One might be either chauvinistic or realistic in assuming women can't stand toe to toe with their male counterparts when it comes to open combat.

Mistress Wing Chun would disagree with that.

Be that as it may, *not* standing toe to toe, unless forced to do so, is a wise decision—no matter if you're male or female—when faced with a superior force—superior in brawn that is, not in brain.

The brain was the battlefield of the Black Lotus. For therein will always lay the strength of such sisterhoods: Men, everywhere, in all times, underestimate women.

Underestimating your opponent is the flip side of a very fatal Janus-faced coin. The reverse being overestimating yourself. Know yourself . . . know your enemy.

Rather than dealing with opponents on a physical level, Black Lotus adepts chose to attack a man where he was weakest—between the legs—which just happens to be controlled by what's between his ears.

(The old joke goes that men are better thinkers than women. Why? Because two heads are better than one!)

But that's just a joke (and not a very good one at that!). The truth remains that we are all susceptible to our emotions,—emotions that can change in the blink—or seductive wink—of an eye.

Ninjo

In Japan, Miko, and other women whose livelihood—and often their very lives!—depended on their ability to please, or at the very least placate, their gentlemen callers, mastered the art of *Ninjo.**

Ninjo means "human emotion," in particular, the emotion of "empathy"—feeling what another is feeling.

---

*Not to be confused with "Ninpo," a synonym for "ninjitsu, the art of the Ninja.

On the one hand this means feeling sympathy for another person, offering them generosity. But on a deeper, darker level, Ninjo refers to feeling *out* what another person is thinking by any means necessary—how they talk, how they walk, their body language.

The art of Ninjo studies all human emotions, and then refines the best ploys and platitudes to either elicit those emotions or short-circuit already existing emotions in others, depending on your agenda.

I know what you're thinking. That sounds a lot like the Ninja's *Gojo-Goyoku,* the "Five Weaknesses," better known at the Black Science Institute as the Five Warning F.L.A.G.S. (Fear, Lust, Anger, Greed, and Sympathy).

And you're right. In any human interaction, emotion is the key.

Remember our discussion on "Hitting Her E-Spot" in *Mind Control* (p. 145)?

Quiet as it's kept, *all* human interaction is predicated on emotion. Figure out how to tap into another person's emotions and you can pretty much run the show. Another way of putting this is "hitting their buttons," or "yankin' their chain."

It's just simple yin yang: If we like someone, if we are attracted to them, we'll want to be around them more, do business with them, perhaps fall in love with them . . . or at least fall into bed with them.

Find someone who doesn't respond to one of the Five Weaknesses and you'll have found the next Gandhi.

But wait. Might we not even be able to entice the Mahatma by yanking his "sympathy" chain? Or how about pressing his *anger* button—"righteous anger" that is, over an injustice done to an innocent?

Everybody's got their emotion "buttons," everybody's got their price. As soon as you go thinking *you're* the exception, that's when you've placed yourself in the worst kind of danger.

They say you catch more flies with honey than with vinegar. But what about bullshit? Flies are attracted to bullshit, too.

In the same, hopefully less crude, way, the tactics and techniques that rely on emotion to seduce can easily be applied to both male and female, in the bedroom or the boardroom.

Gender Plots and Ploys

We're all more monkey than we want to admit.

Try this simple test. Read these pairs of words aloud to your test subject, one pair at a time. Have them assign either "male" or "female" to one word in each pair:

fork-spoon	salt-pepper
pink-purple	cup-glass
turkey-chicken	vanilla-chocolate
truck-car	butterfly-cricket
lake-river	table-bed
New York–Los Angeles	candle-lightbulb
blonde-brunette	red-black

Notice how almost everyone answers the same? That's because we all have hidden gender bias attached to almost everything we do, even to the point of unconsciously assigning gender quality to inanimate objects.

Of course, it could be argued that we've all been socialized to associate the color pink with girls and the color blue for boys, but the truth of the matter is differences between genders go even deeper—womb deep. In other words, certain types of gender differences—strengths and weaknesses—have been wired into us from before we were born.

For some this revelation might be cause for concern, but for the Black Science adept it is cause for celebration since it opens yet another avenue for manipulating and exploiting people . . . uh, I mean, it offers increased opportunities for us to utilize the full potential of our fellow man.

First of all, we all start out with female brains (gasps of horror from all the men in our audience!).

It's only when the developing fetus is dipped like Achilles in the River Testosterone (at about eight weeks) that male and female brains part company—and then spend the rest of their lives trying to get back together! That's why the wiring in the female and the male brain is different.

Some of this we're born with, while other differences either emerge or are exacerbated by social expectations—and/or a combination of the two.

Recent discoveries have pinpointed discernible, obvious, differences in both the anatomy and the chemistry in male and female brains. For example, the female brain grows more connections in the hippocampus (no, that's not a veterinarian college in Africa). The hippocampus is the area responsible for communication, memory formation, and emotion (there's that "E-word" again).

When it's all said and done, her brain will be nearly 10 percent smaller than his, but will have more elaborate connections.

Though having roughly the same number of brain cells, hers will be more tightly packed (because everybody knows women are better at packing!).

Overall, women have 11 percent more neurons (nerve connections) in their brains, especially in the areas that govern language and hearing, than do their male counterparts.

Her communication skills advantage is evidenced by the fact that, on an average day, she'll use 20,000 words, as opposed to his mere 7,000.

The female brain is not only better at talking (Did I *say* bitching? Did I say gossiping?), it also seems to be better at listening (Weise, 2006):

> Being a woman . . . is like having giant, invisible antennae that reach out into the world, constantly aware of the emotions and needs around you. (p. 9D)

More elaborate wiring makes women better multitaskers—or either they are just able to jump back and forth from one job to another faster . . . making it harder for men to keep up.

The male brain, on the other hand, focuses on one difficult task at a time; a man's "inferior" listening ability being compensated for by superior focus.

Whereas her advantage springs from her hippocampus, he has his larger amygdala to thank, that primitive area of the brain responsible for fight or flight, where fear is registered and aggression triggered. Seems he traded cells in his communication centers for more cells in areas of his brain that govern sex and aggression. Yeah, he's a beast.

Nothing emphasizes sex differences more than how differently men and women think about sex—or at least, how often.

The brain space allotted in a woman's head for her sex drive is two and a half times smaller than a man's. As a result, the average woman thinks about sex once a day.

The average man thinks about sex every fifty-eight seconds! (Weise, 2006)

When it comes to the "war between the sexes," men and women also have differing strategies, yet both know enough to downplay their own built-in gender weaknesses, while exploiting the gender weaknesses of their prey.

Woman fact: To hang on to what they decide is Mr. Right, some women will deliberately get pregnant. This is known as a "gotcha pregnancy."

No this isn't a "trailer trash/ghetto rat" stereotype, it's based on a recent study (Bryner, 2005:32).

Man counter-ploy: Make sure the little soldier is always wearing his helmet.

Woman fact: Women have a built-in biological drive to reproduce with men who they feel are good providers (Bryrer, 2005). (Translation: You think she's admiring your buns of steel, but she's really checking out the size of your wallet.)

Man counter-ploy: Show yourself to be a good provider—without too obviously bragging. [Bragging hint: Arrange to have a friend (your "wingman") brag on you, and then you can return the favor when he's trying to score.]

Woman fact: Women are hard wired by Mother Nature to nurture— babies, little puppies, old folks, shy and sensitive men.

Man counter-ploy: Carry around pictures of your children (by your *ex*-wife), pictures of "Grammy," and pictures of your lost puppy. Tell her that story of how you were a lonely boy when you were little. How you secretly like watching movies like *Sleepless in Seattle*, but your male friends just wouldn't understand.

Woman fact: Women pick up on nonverbal cues (body language, facial expressions, tone of voice) much better than men do. Women also *send* nonverbal cues that the average man doesn't pick up on. The average male brain is not skilled at picking up and interpreting nonverbal signals. Men's brains work in a more linear, literal, straight-ahead manner of thinking.

Man counter-ploy: One, stop being "average." Just because the average man has a box of rocks for a brain doesn't mean you have to, too. Learn to control your own body language, your "shadow-walk," in order to give her the signals she's looking for rather than just giving yourself away.

Since women are more sensitive when it comes to picking up on nonverbal cues, learn to send her nonverbal signals that will bypass her rational barriers and attack her subconscious. (More on this in a minute in "The Art of A.W.E.")

Man fact: When men see a problem their instinct is to try to solve it, when they see something stuck, they want to un-stick it. When something's broke (and sometimes when it's not even broke *yet*) they want to fix it.

Woman counter-ploy: Use "The Chastity Belt Ploy," give him a problem in need of solving, something in need of fixing—car trouble, perhaps.

How about a lock, or just the tight lid on a jar you need opened? He'll be glad to help a damsel in distress.

This ploy qualifies as a "sympathy" ploy, with just a little "lust," or at least the promise of same, thrown in for good measure.

Woman fact: Women tend to verbalize a problem, they like to "discuss the implications" before getting their hands dirty. Men, on the other hand, like to jump right in with both feet. (Duh! That's why they call it "tackling" a problem!)

Man counter-ploy: Get really good at pretending to sincerely listen to all her ideas about how something should be done. Make her *feel* she is part of the decision-making process, make her *feel* that you are really considering the problem from her angle—then go right on and do it the way you were going to do it before she started "over-thinking" it to death. *But* . . . you had better be right. So long as your solution solves the problem—without killing any spotted owls or destroying the ozone layer—she'll love you for it. Who can argue with success?

Very important afterthought: If she notices you didn't do it *her* way, be sure to give her a big hug and assure her you only came up with *your* way of doing it properly after you talked to her and she "inspired" you. You will so get laid!

Woman and man fact: We all expect our mates to be interested in everything we're interested in. Not gonna happen.

Woman and man counter-ploy: Pretend to be interested, but don't get caught pretending. Men especially take note: there will be a test on everything the two of you "discuss" later. A test you will fail.

In order for a relationship to last, you both need friends outside of your mates who truly share the same interests as you do. Women shop. Men watch sports. Encourage your mate to have friends. If nothing else, it gives you a needed "time-out."

Pay attention to your mate's needs now, or pay the divorce lawyer—or at least, child support!—later.

Woman ploy: When talking to a man, use simple, short, declaratory sentences. State what you want in outline form. Avoid using too many examples and unnecessary adjectives.

Man ploy: Use plenty of adjectives and anecdotes to attract and engage a woman's attention.

Woman fact: (A hint on how to win an argument.) Since women are better at remembering, expect that during an argument she's gonna bring up something from the past.

Man counter-ploy: You can only fight one war at a time. And since the male brain finds it easier to forget (or at least suppress) emotional situations and move on, you probably don't even remember the incident she's dredged up from God knows how many years ago. Instead of trying to counter her ancient allegation—even if you do remember it, try this: first compliment her fine memory, before then bringing her back to the present problem. Don't try to brush away her "examples" from the past, that will only make it look like you're lying or refusing to 'fess up. Tell her:

> Honey, I know that since you can remember things from the past so well [compliment] that you'll also remember that every time we've had a problem, every time something like this has come up in the past, we've always managed to get through it . . . together. And this time will be no different.

Can you say make-up sex?

The Art of A.W.E.

> *"Male dominance depends on female awe."*
> —**Jesse Bernard,** *The Sex Game.* **Prentice-Hall, 1968:59**

Admittedly, the sections on the Art of Seduction in both *Mind Manipulation* (2002) and *Mind Control* (2006) took a decidedly male perspective. This is in no way meant to imply that only women are susceptible to the plots and ploys of seduction. Quite the contrary.

One could argue that men are actually more susceptible to seduction since they are so darn-awful sure they can't be seduced.

It's the old argument that if you perceive something as a threat you arm yourself against it. But if you don't see something as a threat—like someone actively trying to seduce you—why bother building up your defenses? Besides, when most men think of being "seduced" it usually involves their ending up in bed with a beautiful nympho. In other words, men have a hard time seeing the downside.

Men are natural-born predators (in a good way) and just don't see themselves as "prey," though, for some reason, they all think of themselves as a good catch!

Black Lotus operatives, Ninja kuniochi, and other wily women are quick to exploit this contradictory thinking, other crackpot ideas, and cracks in a man's mental and emotional armor.

Remember, "Two heads are not better than one," especially when they can't agree as to what's the best place to spend the night.

Man ploy, woman ploy, doesn't matter. The *principles* inherent in any gender ploy can almost always be turned around, and turned to your advantage.

Remember the promises of Shock and Awe during the battle for Baghdad?

"Awe" is defined as "a feeling of reverence and dread, mixed with wonder" (*Webster's II New Riverside Dictionary,* 1996).

In Black Science, particularly when studying the Art of Seduction, "awe" can be applied to any human interaction in which you (1) want to make a good first impression and (2) overwhelm them with your charm.

There is both a conscious level and an unconscious level to *awe*ing someone (see figure 17).

Conscious-level *awe* strategy is a three-step procedure, designed to bring you closer to your goal, your "end game."

Thus, step two is dependent on your receiving the right feedback to your overtures in step one. Likewise, proceeding to step three depends on how well you do—how well you "woo"—at step two.

At step one you test the waters, you try to make a good first impression. It doesn't matter if you're trying to catch the eye of some hot babe in a bar, or trying to get your foot in the door to sell vacuum cleaners.

It's important to note that, at this conscious level, you make it clear to your "target" audience what it is you're trying to sell—whether yourself or handy household conveniences.

At this conscious level, "A" stands for "approach," i.e., how you make first contact with your target (whether for sex or sales).

"Approach" also refers to your "sales pitch," your initial spiel that you will embellish upon in the next step—if you make it to the next step.

If your overtures are ignored or even rebuffed at this first step, you simply move on to a better fishing hole where you keep casting your line out until you feel a nibble on the end of your pole.

Step two, "W," stands for "wooing." To "woo" means "to court," "to entice and entreat intently."

That babe at the bar didn't mace you. You didn't get the door slammed

"THE ART OF A.W.E."

	A	W	E
	"APPROACH"	"WOO"	"END GAME"
I. CONSCIOUS LEVEL	Testing the waters. Initial overtures. FIRST IMPRESSIONS. Introduce your "product".	Show your wares. Court, entice, entreat. Listen and respond, adjusting sales pitch to fit needs of consumer. Close in on sale/committment . . .	Close the deal. Get paid. Whether selling a used car or selling yourself, here is the payoff. Get them to sign on the dotted line
	POSITIVE FEEDBACK >	POSITIVE FEEDBACK >	
	A	W	E
	"ANGLE & AGENDA	"WEDGE"	"END GAME THE SAME"
II. UNCONSCIOUS LEVEL	Hidden agenda. Your target doesn't know what you're REALLY selling. Pique their interest. Appear non-threatening.	Increase the pressure. Drive a "wedge" between your target and anything or anybody that might block your "sale" (e.g., her girlfriends, her beLIEfs, etc.)	Get sex. Money. They join your cult, etc.
	POSITIVE FEEDBACK >	POSITIVE FEEDBACK >	
	(preferred, not mandatory)		

Figure 17.

147

on your foot in your debut as a vacuum cleaner salesman, so there's still a chance for a sale. Therefore, in both instances you've just been given a golden opportunity to present your case, to show your wares and highlight what you're selling in the hopes your audience is in the mood to do some buying.

Or perhaps he or she is undecided as to whether or not they are "in the mood" to do some buying. This is your one chance to convince them there's never a better time to invest (in you!) than right now.

Don't just tell your target about yourself (or about the product you're trying to peddle), listen to them talk, listen for a clue to what they *think* are their needs, what they *think* they are looking for, or need in their life.

Is she looking for a husband . . . or just for a good time? Is she looking to buy a vacuum cleaner . . . or just looking for a good time!

In whatever business you're in, it's all about making the sale, and making the sale is all about selling yourself.

The "E" at the conscious level of *awe* strategy stands for "end game." A successful completion to your wooing attempt.

Unconscious-level *awe* strategy takes on a more covertly focused—possibly felonious—dimension.

The unconscious level is one continuing step. Here "A" represents "angle and agenda."

Unlike the slow, careful "feeling out" process at the conscious level "A" step, where it's obvious to your target what it is you're trying to sell, here at the unconscious level, you don't let the other person know what it is exactly you're selling or that you're trying to get from them—sex, money, unquestioning obedience to your cult.

You already know your goal, your agenda, and you have already mapped out your "angle of attack" to help you (ruthlessly) reach that goal.

Here, rather than respectfully testing the waters with your big toe, you jump right in with both feet! You hit the ground running with your "endgame" goal clearly in mind. It doesn't matter if that end game is simply gettin' some end, scamming some poor sucker out of his life savings, or something even more sinister (hey, you're from Texas, you own a chainsaw and a hockey mask . . . people judge!).

"W" at this unconscious level is the "wedge" you have to drive between her and her "cock-blockin'" friends, in order to get her alone long enough to shoot your line of bull. Or you may need to drive a wedge between her and her natural suspicions, in order to get to her pocketbook.

More sinister still, you might need to drive a wedge between a person and their current beliefs in order to get them to "just check out" your new cult.

When trying to "brainwash" captured POWs, the first step in breaking down the prisoner is to drive a wedge between him and his fellow POWs— sowing suspicion between them, e.g., by giving one special privileges or position over the others (Lung, 2003).

We drive a wedge by planting doubt, in effect planting a question in our target's mind—a question we already have the solution for.

The "E" at this deeper, darker unconscious level is the same as at the conscious level—"end game." The major difference is that, at the conscious level, both you and your target know what you're aiming for, whereas at this unconscious level your ultimate end game might be to strip her of her resistance to a one-night stand, rape her ATM, or molest her mind inside your cult compound.

Some *awe*ing tips:

Look for the "Look": Look for her head cock. (No, not *that!*)

Look for the way she cocks her head to one side, resting it on her shoulder, in effect exposing her neck to your vampire lust. Her body language is telling you she trusts you.

She looks at you and then quickly looks away. Bummer! But wait, notice how her head is still pointed in your direction; more body language showing she's interested, she's inviting.

And notice how that quick smile lingers on her face even when she looks away, betraying her pleasure in seeing you. Remember: a real smile shows in the eyes as well as on the lips. (See figure 18, next page.)

Sexual Feng Shui: Don't let anything come between you and her—literally.

Always try and catch her away from her potentially "cock-blockin'" girlfriends. The way the King of Beasts culls a single antelope from the herd!

Meet her in a side-by-side position (at a bar, buffet table, or salad bar); that way she'll "turn into you" in order to stand face to face, a certain sign of interest.

When sitting, choose small round tables, rather than larger, square-edged tables.

Don't hold your drink, buffet plate, hat, briefcase, or anything else between you and her. We're trying to tear down walls here, not build more.

We want open line-of-sight at all times—like a sniper. Our center line

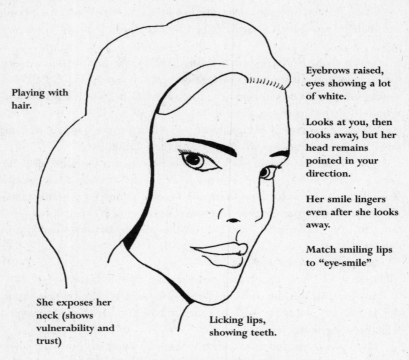

Playing with hair.

Eyebrows raised, eyes showing a lot of white.

Looks at you, then looks away, but her head remains pointed in your direction.

Her smile lingers even after she looks away.

Match smiling lips to "eye-smile"

She exposes her neck (shows vulnerability and trust)

Licking lips, showing teeth.

Figure 18. Sure signs she's interested in you.

(imaginary line running from your face to your groin—ironic, huh?) should be lined up with her center-line as much as possible. Keep your right hand empty, open, and inviting to shake.

"Mirror" her body movements, but don't mock. Match her physical tone, i.e., if she's animated, be animated, too. If she's sedate, assume an air of calm about yourself.

Be Eager, Not Aggressive: Wait for her to lead. Follow enthusiastically.

Avoid aggressive gestures and postures. For example, hand on your cocked hip or dangling near your crotch says, "I'm a gun-slinger who keeps his hand near his six-shooter at all times." Legs wide apart—you're advertising. Leaning too far forward—"Come on, let's wrestle!"

What About Gender-Benders?: In Japanese Kabuki, men play all the roles, including the female parts—as it was in Shakespeare's day. Masters of this feminine impersonation are called *Onnagata,* and are much honored in Japan. The argument being that it doesn't take much for a male actor to play a man on stage, but it takes one helluva male actor to play a female convincingly.

So impressive was (and still is) the Onnagata art that, during medieval times, young Ninja were often sent to study with such masters in anticipation of assignments where they might have to dress in drag in order to get close to a targeted Daimyo. (See the section on "The Art of Disguise" in *The Nine Halls Of Death*. Lung & Tucker, Citadel, 2007.)

Then there's the story of the British diplomat to Communist China who fell in love with a beautiful Chinese woman, who subsequently gave birth to his son. The three of them lived very happily together until he was ordered to return to Britain.

In order to safeguard his wife and son still back in China, the diplomat was forced to become a spy for the Chinese. Only after he was finally arrested by his own government for spying did he discover the truth: his "wife" was really a man, an accomplished Chinese transvestite!

We've all heard our buddy's excuses: "It was really dark in the bar . . . and, uh, I was really drunk. . . . No, I didn't notice her Adam's Apple. . . . Well, you know some women just have naturally deep voices. . . . Anyway, by then I'd already paid for the room so I thought, what the hell!"

Yeah. In this day and age it's easy to get fooled . . . just ask Eddie Murphy.

That's why you always "measure twice, cut once."

But before such a surprise encounter leaves you with a bad taste in your mouth, you should realize that the same seduction strategies, tactics, and techniques that work male-on-female, female-on-male also work on gays and lesbians.

More important, whether for your personal protection, or some private predilection you'd rather not talk about, keep in mind that "gender-benders" can use these techniques, too . . .

Happy hunting. Watch your six!

"Within every man there is the reflection of a woman, and within every woman there is the reflection of a man."
—Hyemeyohsts Storm

SEVEN SINISTER SISTERS

While we're on the subject of women being "the deadlier of the species," we shouldn't forget the "Seven Sinister Sisters," a cautionary lesson taught by the infamous fraternity of International Finders—actually a thinly disguised road map for causing our enemy to "misstep."

Seven Sinister Sisters

Misses   **Hits**

- *Misinformation:* Feed him false information. Plant the seeds or doubt and distrust in his garden. Make him mistrust something or someone he needs to trust within his immediate circle. Read *Othello.* Make him want (need) to investigate. Get him up off his ass and moving . . . in the *wrong* direction! Misinformation gets her cousins Misdirect and Mislead to help.
- *Misperception:* Encourage him in his investigations. He sees what *you* want him to see, he finds what you strategically place for him to find. He sees a photo of his girlfriend laughing with a strange man (a man you hired to "accidentally" bump into her on several occasions).
- *Mistrust:* He confides his suspicions to you. You make a show of "objectively" examining his evidence before coming to the "logical" conclusion that he's come to a logical conclusion in his suspicions. But, you caution, he should keep his suspicions (and conclusions) to himself for the time being. After all, others might be in on this "conspiracy" against him. *You* are the only one he can truly trust. He can't even trust himself and should therefore give over more and more of the burden of running his business, etc., to you.
- *Misfortune:* As if things weren't bad enough, now little things seem to be going wrong in his life. He might be able to deal with the "major situation" if not for all these little problems suddenly popping up all over the place. Like the true friend you are, you offer to shoulder more and more of his business and financial burdens . . . even as you busy yourself creating even more "little problems" in his life—"cutting at his edges," siphoning off his resources and energies.
- *Misdemeanor:* He's nearing the end of his rope—and it continues to fray!— and you "suggest" that perhaps desperate times call for desperate

Misses   **Hits**

measures; that some thoughts *and actions* he wouldn't have dreamed of contemplating before, might now be "justified." Perhaps you talk him into stalking that girlfriend he believes is cheating, or rifling through his boss's files looking for "evidence" the boss intends to fire him. Perhaps you convince him to commit a "small" felony to get that grub-stake he needs to get "back on top." Tell him not to worry, you'll drive the getaway car. You—his only friend—are the only one that will know. Heh-heh-heh.

- *Mishap:* His mistrust (of girlfriend, boss, etc.) is revealed. That "little" and "justified" misdemeanor performed in the desperation of night suddenly seems so much more dastardly when revealed by the glaring light of day. People's faith in him is shattered, revealed by some small *miss*-calculation, suddenly exposed due to some slight flaw in his master plan . . . a flaw you surreptitiously implanted while helping him craft his plan!

- *Missing:* He goes "missing"—shunned by his former friends, perhaps even by family. Fired from his job. Voted off the island. Worst-case scenario: people really do go "missing," disappearing off the face of the earth (although they're usually only six feet beneath the surface of the earth!), especially after having started "acting strangely." People often go "missing" right before they have a change of heart, repent, and decide to tell all—implementing their co-conspirator . . . you! Read Shakespeare's *Much Ado About Nothing.*

Misinformation

The cook who stirs the pot, with the help of her two equally treacherous cousins, Misdirect and Mislead.

The more bull we feed our foe the more he'll suffer from diarrhea of the mouth and constipation of the brain. In the end, he'll cook his own goose!

Misperception

Smoke and mirrors. We make him see what isn't there, and ignore what is. Having learned how easily the eye and other senses can be tricked, we use this knowledge to distort his perception of reality. The more our enemy's reality diverges from the boss's reality, the better for us to overcome him.

Mistrust

Now mistrust sets in, when our enemy realizes he can't trust his own perceptions, can't trust himself. Such a realization can lead him to become despondent, make him want to give up . . . or else come to us for help. Thus, when crafting such a scenario, when herding your victim down the path of doubt, self-loathing, and despondency, remember Sun Tzu's advice to always leave a fleeing foe a way out . . . straight into your arms!

Misfortune

We undermine his confidence by convincing him kindly Fortune has turned her back on him and her evil twin Misfortune has set her seductive sights on him.

Little things begin to go wrong in his life. His car won't start so he's late for work. His secretary calls off because her son got caught with a baggie of marijuana—little stuff that collectively ruins his day. Ruined days add up to ruined weeks, ruined months and years, and lives.

Musashi: We cut at his edges; little irritations at first, minor inconveniences that become bigger and more unmanageable as they begin to pile up.

This is pure guerrilla warfare. We *make* him sweat the small stuff, draining his energies, making his boss wonder, "If he can't handle a little thing like that, can I really trust him with that major account?"

Distracted with all the little things going wrong in his life, he misses the big picture unfolding right in front of his eyes.

Misfortune besets him, misfortunes real . . . and those covertly crafted by you. To accomplish this, you may need to employ Misfortune's cousin Miscellany—attacking him with a mixture of unlikely and unexpected events, a mix-n-match of odd occurrences and highly improbable accidents.

Misdemeanor

His confidence shattered, finding himself betrayed by Fortune, we now urge him in the direction we want him to go by getting him to "misbehave," to do something—just a small thing—something against his nature.

One small change in behavior leads to big changes later on. Minor indiscretions, faux pas that lead to felonies. Little things we can later hold over his head in a big way.

We get him to "go along to get along." We make him act out of character. Remember, this is the first step to brainwashing POWs in wartime. Convincing them to do "this one little thing" that, once done, makes it a whole lot easier to do it the next time, each time taking another step toward total obedience.

Before long, that single small step (literally, a step out of character) leads to his committing more serious breeches, ultimately resulting in our "banking" bloodties that guarantee his future cooperation.

Mishap

We build on his mistrust, and on the misfortune that has beset him. We befriend him and he comes running to us for shelter when mishap comes hunting him. Trouble now has him on speed-dial.

He's scared to come out of the house. People have begun to shun him—bad things just seem to happen when he's around. Former friends begin to see him as a jinx, a pariah. Before long he's seeing himself the same way. You're his only friend.

One day there's an anonymous caller at the other end of the phone asking how much he's willing to pay for the photos of him and that seventeen-year-old hooker. The wrong word gets dropped in the right ear and a new job just opened up for your brother-in-law.

Good thing you're right by his side to help him weather the storm . . . almost like you can somehow predict the weather!

Missing

Worst-case scenario and none of the first six sisters makes a dent in your enemy's determination to oppose you, there's still that most "permanent" of the Seven Sinister Sisters—Missing—who just happens to coincide with the last of the "Six Killer B's."

Maybe your enemy goes missing because he's suddenly come under investigation for some heinous crime . . . maybe skipped out to avoid some mysterious debt . . . maybe he ran off with a *mis*tress no one knew about.

Long as there's a good—or at least reasonable—reason he disappears nobody—mainly John Law—is going to suspect foul play.

People disappear all the time . . . unfortunately, not always the right people.

NINE LADIES DANCING

From 1558 to 1829 Roman Catholics in England were not permitted to practice their faith openly. According to popular lore, Jesuit priests supposedly wrote "The 12 Days of Christmas," hiding instructions from the Catholic Bible inside the lyrics.

FYI: The antics of Jesuits, aka the Society of Jesus, founded by Spanish ecclesiastic Ignatius of Loyola (1491–1556), bear looking into on so many Black Science levels.

At one time or another, the Jesuits have been accused of acting as both Stormtroopers and Gestapo for the Vatican.

To their credit, Jesuit missionaries were fearless (some say "ruthless") in traveling to far-distant lands in their effort to make sure native women everywhere were wearinq European-made brassieres.

In Asia, the Jesuits doffed their sinister black robes in favor of saffron robes—so they would better be confused for Buddhist priests and more readily welcomed into the homes of potential converts.

It was an eighteenth-century Jesuit stationed in China who translated the first Western version of Sun Tzu's *Ping-Fa* into French.

But I digress. "A partridge in a pear tree" refers to Jesus, while "twelve drummers drumming" is code for the twelve points of the Catholic "Apostle's Creed."

Of special interest, Black Science–wise, are the "nine ladies dancing," since this phrase secretly lists "the nine ways in which one can be an accessory to another's sin."

Of course, when it comes to plying the Black Science, being "an accessory to another's sin" isn't necessarily a bad thing. Quite the contrary. By reading these nine no-no's from a Black Science perspective, we discover nine new avenues down which we can travel to further distract and detour our enemies:

Nine Ladies Dancing

- *By counsel:* We provide our foes with false information. We give them "The Mushroom Treatment": We keep them in the dark and feed them bullshit!

No matter how hare-brained their scheme, no matter how bizarre the "conspiracy theory," we assure them they have come to a "logical" conclusion and should act on what they "know" to be the truth.

- *By command:* Once we have von their trust and/or whenever they are otherwise under our control and we are "in authority," we send them on "suicide missions." These are what Sun Tzu called "expendable agents."

- *By consent:* We give them our blessing—either overtly, or through promoting an atmosphere of "permissiveness."

- *By provocation:* We "yank their chain," we stir up their emotions. We implant doubt and greed and other negative emotions to cause the person to act irrationally and/or ally themselves with our agenda.

- *By praise or flattery:* We inflate their ego until we can pull it along behind us like a balloon on a string.

- *By concealment:* We use lies of omission, we tell half-truths, we pass rumor and innuendo and propaganda off as fact. Like any good cult, we control the flow of information. We also help him conceal damaging information of his indiscretions . . . always keep a backup disk!

- *By sharing:* We become his partner-in-conspiracy. We gossip with him. We keep his secrets . . . all "bloodties" for later use.

- *By silence:* Omerta. We often do more by our *inaction* than by our actions. Our "silence" buys us trust . . . which we are free to abuse at any time!

- *By defending a wrong that has been committed:* We co-sign for a glaring wrong/crime/sin he has committed; helping him justify it in his mind. He becomes dependent on us for his feeling of self-worth. By co-signing for a small wrong he has done, we can then talk him into committing an even greater wrong (perhaps in an effort to rectify the first wrong).

By Our Counsel

"The sad truth is that most evil is done by people who never make up their minds to be either good or evil."
—Hannah Arendt

Our advice to him is a vice to him; a vise with which to seize and hold his attention, an amoral vice with which to tempt, addict and weaken his resolve and resistance.

We feed him misinformation, rumor, gossip, and propaganda. Besotted with this Bordeaux of bullshit, drunk and disoriented from the liberal amount of disinformation we've poured his way, he cannot help but stagger, stumble, and surely fall into mistakes from which we can profit.

By Our Command

When in a position of power or authority over him, we can easily send him on assignments that spell his doom. Remember how "good" King David sent loyal subject Uriah to his doom just so he could steal Uriah's wife?

But even when our rival is on an equal footing, we can still give him "authority" by helping him rationalize his behavior, in effect co-signing for him.

We can also convince him God is on his side, that his cause is not just "just," but is just what God ordered. With God as your co-pilot (or in this case, co-signer), what other "authority" do you need? Just ask the Taliban!

By Our Consent

We give our blessing to him, agree with his convoluted reasoning, his excuse for doing whatever stupid thing he's already decided to do anyway. We encourage his extremism.

By Our Provocation

We "yank his chain." When he's mad, we go out of our way to enrage him further. We encourage him to seek revenge for slights against his person real or imagined . . . or, better yet, manufactured by us.

Self-importance is our greatest enemy. Think about it—what weakens us is feeling offended by the deeds and misdeeds of our fellowmen. Our self-importance requires that we spend most of our lives offended by someone. (Carlos Castaneda)

Show him those pictures of his wife slipping around behind his back.

We egg him on, "Are you really gonna take that? Stand up, act like a man! Don't let that guy punk you out like that!"

By Our Praise or Flattery

Tell him how brave and smart he is to have decided to go down that (hopefully disastrous) path he's chosen.

By Concealment

Help him hide his indiscretions; "banking" them for a time you will be able to use them against him.

If he's waffling and wavering, thinking about throwing himself on the mercy of his wife, his boss, the police, confessing his indiscretion—talk him out of it!

Encourage him to keep it to himself. (So you'll have a "bloodtie" to hold over his head—a common ploy with gangs and cults.)

Encourage him to lies of omission. Don't ask, don't tell. Convince him that keeping secrets from those he really cares about, he's doing them a favor, protecting them from bad and embarrassing news.

By Sharing

Join him in his conspiracy. Where needed create a conspiracy and include him in it before he has a chance to decline your offer to join.

Give him a piece of information that implicates him just by his knowing. The law calls this "accessory after the fact." Insider trading. Paging Martha Stewart.

By Our Silence

The Mafia calls it "Omerta." Make him swear an oath, swearing on his honor and the life of his favorite dog.

Make yourself his bosom buddy, his only confidant, the keeper of his sacred secrets . . . all bloodties you can "bank" for later.

By Defending Wrong That Has Been Committed

Whatever he's done wrong, whatever you've *helped* him do wrong, assure him it was justified. He did the right thing. Be his "enabler."

Make him feel better about the evil he's done (for you) and he'll do even more evil (for you).

5.

Japan: Silk and Steel

"The heroes of ancient Japan love and die within their shells of silk and steel."
—Yourcenar, 1986:7

The history of Japan is one of beauty and blood. Beware: Blood can be just as seductive as beauty.

The sword has always held an honored place in the heart of the Japanese people. Indeed, the sword is one of Japan's sacred "Three Treasures" (see below).

The Japanese kanji *"bu"* means "sword." Hence a *"bu-shi"* is someone carrying a sword, a warrior. But "bu" can also mean "pen."

Samurai were expected to be both men of action, as well as patrons of the arts. Thus the ideal: *"Bunbu itchi,"* to have *"pen and sword in accord."*

We find this ideal exemplified throughout Japan's violent history, not just up through their middle ages with the warrior-author Miyamoto Musashi, but further on to World War II *kamakaze* taking time out of their busy dying schedule to calmly write their death poems, down to modern-day mystery man Yukio Mishima, the famed Japanese author and actor who, at the head of his Samurai militia, committed *hara-kiri* ritual suicide—in 1970!

Thus the Samurai spirit is without question worthy of our time and effort to study: from Musashi's masterpiece of strategy, *A Book of Five Rings,* down to the study—and hopefully mastery—of the mind-set, discipline, and focus it would take to pull off "a Mishima."

But it is not just Samurai we will study during our Black Science sojourn in Japan. There are other beautiful—and, yes, often bloody!—Japanese cadre worthy, if not of our admiration, at least of our attention.

And while they are all guilty to some degree of using similar methods of mind control to discipline themselves, and then mind manipulation to corner and then convert or coffin-up their foes, we may find they differed greatly in what motivated them.

Different internal energies—and sometimes external forces—act on and animate the shining Samurai, than do the shadowy Yakuza gangsters, and the downright dark and deadly Ninja.

And while it's true all these groups are bathed in both beauty and blood, there is still much we can learn from their physical and mental disciplines— disciplines that allowed them to first conquer themselves and then overcome their enemies.

Ours is not to pass judgment on their attitudes and actions with pounding gavel but, rather, to dispassionately observe and note with purposeful strokes of our pen and, if need be, with just *as* purposeful strokes of our own sword. *Bunbu itchi!*

THE THREE TREASURES

Tenno-rei, the "Imperial Soul," is the Shinto belief that each new Japanese emperor receives at coronation a sort of cumulative, eternal "consciousness-soul" that has, in turn, passed down through the entire historical line of emperors, back to the divine origin of the first emperor of Japan.

Also passed down from emperor to emperor are the *Sanshu no shinki,* the "Three Treasures."

According to myth and tradition, the Three Treasures were given the Japanese Imperial line by *Amaterasu,* deity-creator of the Japanese archipelago.

These Three Treasures are the sacred sword, the sacred mirror, and a sacred string of jewels.

But the Sanshu no shinki are not just physical objects.

It is believed these sacred objects exist simultaneously in three worlds: this physical world, the spiritual world that overshadows the physical, and the world of ideas (thought) that bridges the first two.

As physical objects they are undoubtedly real and, de facto, exist in the physical world. As sacred spiritual objects they hold a special place in the mythology and ritual of the Shinto religion and represent the Imperial identity.

But there is also a mental level lesson to be learned from these three, a Black Science lesson.

As often as the future fate of Japan was decided on fields of bloody battle, with a shimmering stroke of samurai sword (or perhaps a Ninja blade to the back!), just as often lives were lost and an empire won due to intrigue within the Imperial Palace itself, where rich and powerful Samurai clans vied for the ear of the emperor—all the while plotting to replace him with one of their own.

"The *shoji* have eyes" has been a common saying since the founding of the Imperial Court.

Shoji are those sliding doors you find in traditional Japanese homes, opaque paper "windows" on wooden frames. A master intriguer was sometimes referred to as a "wet finger," alluding to the fact that one only had to wet a finger to poke an eyehole through this thin paper veneer in order to spy on whoever was on the other side.

In the same way the Japanese "borrowed" much of their culture (writing, religion, martial arts, etc.) from China, so too, it seems they borrowed many of their more masterful—albeit still despicable—techniques of court intrigue from their Chinese cousins.

China has never been a stranger to, nor stingy with, the art of intrigue.

Sun Tzu, in his *Ping-Fa* chapter on spies, speaks of "turning" disaffected bureaucrats and cunning courtiers to your cause.

Sun Tzu wrote in the fifth century BCE but, as early as the Chou Period (twelfth century to seventh century BCE), in the first anthology of Chinese poetry, the *Shin Ching* ("Book of Songs") we find a now familiar saying, and a warning:

You must guard well your idle words, for the walls have ears.

Strategist Kuan-Tzu (fourth–third century BCE) echoes this same warning:

In olden times we were warned "The walls have ears." This refers to vital intelligence overheard by someone other than The Emperor for whom that information was intended. One way this disaster comes about is through the wiles of the sly courtesan who uses her charms to lull The Emperor into a lax mood, where she can then tickle such secrets from him. In the end, only her evil pimp benefits.

Of course, one need only look at the history of Rome, study Machiavelli, or simply read the Bible for that matter, to be able to convincingly argue that Asian Imperial courts have never had a monopoly when it comes to back-biting and back-stabbing.

Whatever the shape of the key hole, there is sure to be an eye to fit it!

The word *"Mekura"* literally means blind but refers more to those whose eyes are fine but who lack an "inner eye" for subtle and even esoteric understanding, particularly when it comes to intrigue.

To get ahead—or at least to keep the head you had!—all the while scheming to get the political upperhand against the other hungry houses and clans, meant you had to develop your mekura inner eye. But, since deliberate court intriguing, if caught, could easily cost you—and your whole family!—their head, wily Japanese courtiers never spoke openly of such things as covert surveillance, conspiracy, and corruption. Instead they used overly polite and subtle euphemisms—a kind of covert Court code.

Often when plotting they used euphemisms taken from Noh theater (see section below).

Other times, intriguers used the Three Treasures.

Use of the "Three Treasures Strategy" gave intriguers the option of several tactics of pushing their enemy over the edge into the abyss; pulling them into a trap; or simply using numerous ploys to ensnare their Court rivals.

And, as you've no doubt already figured out, these Three Treasures are easily comparable to the Five Warning F.L.A.G.S. you're already familiar with (see page 165.)

The Sword Approach

The sword is truly a universal symbol. In the West, the sword has carved its way into myth and history, representing the power and (hopefully) wisdom of kings—King Arthur's Excalibur for example. It has also come to represent the courage and determination of the solitary hero seeking justice—the sword of Sigurd and Siegfried.

It is the same in the East, and nowhere more so than Japan where the sword is thought of as the "soul" (kami) of the Samurai.

SANSHU NO SHINKI-JUTSU
"Three Treasures Strategy"

Treasure	Approach	Symbolic of . . .	Confucian Interpretation	Strategy	Five Warning F.L.A.G.S.
SWORD (yang) (exoteric)	PUSH	Sword of truth, justice, karma	Wisdom and courage	Repulse and threaten, show willingness to use force.	Fear and anger
JEWEL (yin) (esoteric)	PULL	Reincarnation, chakras aligned wealth	Benevolence	Attract, bribe, dazzle, and confuse	Lust and greed
MIRROR (balanced)	PLOY	Pure soul, Buddhist ideal	Honesty, thankfulness	Reflect his attitude (Judo), stifle, appeal to his altrusim.	Sympathy

Figure 19.

Various legends are told of the origins of the Three Treasures, beyond their having been given originally by the gods.

When it comes to the sacred sword, one story tells that *Susano* the Storm-God slew a great dragon and found the sword in its tail.*

In China, Confucian philosophy sees the sword as the symbol of wisdom and courage.

Musashi's *A Book of Five Rings*, while overtly an actual sword fighting opus, has subsequently been (re)interpreted to reveal a more covert—metaphorical—and cerebral lesson.

The sword defends, it repulses the enemy. But all good swords cut both ways.

Didn't Alexander the Great imprudently and impatiently use his sword to slice through the Gordian Knot? Rather than enrich himself, and prove himself worthy, by unraveling the riddle by hand?

---

*The Shinobi Ninja clans of medieval Japan trace themselves back to this same Storm-God. See Lung, *Ninja Craft*, Alpha Publication, 1997.

The power of the sword is *fear* and all too often swords are wielded in *anger*. Recall that these are two of the dreaded Five Warning F.L.A.G.S. that can all too easily weaken us to our enemies.

In the West, Ms. Blind Justice often holds a sword, a sword that quickly cuts to the heart of the matter—to the truth. Sometimes the sword is then called on to mete out justice and retribution—or *karma*, as it's called in the East.

Using the Three Treasures' "sword approach," we confront our enemies head on. We *push*, and we keep pushing until he bends and ultimately breaks.

We use our sword (intelligence and force) to instill *fear* and *anger* in him. Filled with fear, he will be paralyzed to raise his own sword against us. Or, just as deadly, we anger him into making rash moves, ill-advised maneuvers that place him at our mercy.

No story better illustrates this than that of the "47 Ronin" whose master was goaded into anger by his enemy into drawing his sword while visiting the Imperial Palace.

Drawing your sword on Imperial grounds was considered a threat and affront to the emperor himself and merited the death penalty.

As a result of his allowing his anger to get the best of him, of having fallen prey to his rival's sword/anger strategy, the Lord of the 47 Ronin was ordered to commit ritual suicide.

The 47 Ronin ultimately succeeded in getting their revenge by using a Mirror Approach ploy.*

The Jewel Approach

A jewel-based strategy uses the opposite tact than does a sword-based approach. By its very nature a jewel dazzles us, a hypnotizing gem that draws us inexorably to it.

Thus, rather than *push* like the *sword*, the *jewel pulls*, like a spider attracting us deeper and deeper into its web until escape is impossible.

A strategist employing jewel-based thinking (a "*jewel*er," if you will) seeks to ally himself with those of equal or greater power; an alliance guaranteed to (1) immediately strengthen his own prospects and/or increase his profits

---

*For a more detailed accounting of the 47 Ronin, see Lung and Prowant, *Mind Manipulation*, Citadel 2002.

while (2) setting the stage for his eventual dominance. Can you say "Machiavellian"?

This kind of crafty covenant is preferable to having to play the "*swordsman*" and go toe to toe with an enemy you're not sure you can defeat in open battle, perhaps even knowing you're not yet strong enough to match him move for move, blow for blow.

Better to bond with him than bleed for him.

Since our jeweler is trying to attract (*pull*) his target closer, he freely resorts to seduction and bribery.

You'll no doubt recall our buddy "bribery" as one of the dreaded "Six Killer B's"? (*Mind Control*, 2006)

Jewelers invented "strange bedfellows" and "marriages of convenience." Indeed, a person dominated by this type of thinking would have no qualms about marrying for the money.

Confucian philosophy associates the jewel with the positive attributes of benevolence and wealth.

Symbolically, jewels have been used in the East to represent reincarnation (i.e., many lives, like gems hanging on the single string of the soul). Jewels have also been used to represent the chakras within the human body, physical and psychic power centers activated once our innate, but usually dormant, kundalini energy is awakened through the practices of yoga, meditation, Tantric sex, and other physio-mystical disciplines.

On a more negative, albeit still useful, note, jewel strategy corresponds to both *lust* and *greed* from the Five Warning F.L.A.G.S., since both these are predicated on want and desire, i.e., attraction.

The Mirror Approach

When you can't overtly *push* (sword) or *pull* (jewel) an enemy into seeing things your way, you have to resort to a more cunning method, a *ploy*, a *mirror* strategy.

Using *mirror* strategy, we deflect criticism and discourage scrutiny—why bother with me, I'm so unimportant, so harmless.

This is Sun Tzu 101: "When close, appear far. When strong, appear weak," all in order to cause your enemy to underestimate you, seeing you as no threat. (Read Robert Graves's *I Claudius*.)

In modern parlance this is called "rocking him to sleep," i.e., getting your enemy so relaxed he never sees your rock coming!

This approach also teaches us to "reflect" our enemy's attitude (duh! Like a mirror?), pretending to agree with him, perhaps stalling him with negotiation, until the time is ripe to show our true face.

This strategy is the essence of the spy's craft, the heart of the prostitute's art, and part and parcel of the professional politician's pile of . . . polemic.

We feign weakness in order to get our enemy to drop his guard.

Their lord having been condemned to commit *seppuku* after having been brought down by his enemy's *sword/anger* tactic, it was expected the condemned man's forty-seven Samurai retainers would likewise commit suicide, as was customary.

Instead these forty-seven Samurai—accomplished warriors who had each proven themselves time and again in fearless combat—decided to disband and to suffer the insults of their fellow Samurai for having chosen to live as dishonored "ronin" (masterless Samurai) rather than do the honorable thing and follow their master into the void.

. . . Or so it seemed.

Realizing they couldn't hope to succeed in revenging their dead Lord by taking on their enemy's superior numbers right then and there by employing a direct sword strategy, the 47 Ronin adopted a *mirror* approach. Pretending weakness and cowardice, biding their time until the offending Lord had long forgotten them, only then did they secretly reassemble and successfully storm their enemy's castle.

The next morning, all 47 Ronin knelt over their Lord's grave and, after placing the head of his enemy on that grave, all forty-seven then committed ritual suicide.

A true story.

Mirror ploys require the greatest of weapons: patience.

In the West there is the story of how King David's son Absalom, unable to take immediate revenge against his brother Amnon for raping their sister Tamar, bided his time for two years. Finally, Absalom invited his brother to a banquet where the doors were then locked. After Amnon got drunk on wine and feasted his fill, Absalom was finally able to feed this sexual predator his just deserts! (See 2 Samuel: 13.)

Chinese *Tongs* were infamous for inviting targeted guests to banquets, only to poison them (cf. Seagrave, 1985).

A truly marvelous *mirror* patience ploy involved Chinese feeding specially bred silk worms poison and then using their toxic silk to make exquisite robes that would then be given as gifts to one's enemies. After the robes were worn for a while, the poison would slowly leech out of the fabric and into the wearer's skin, sickening and/or killing them (cf. Lung and Prowant, 2001).

Recap: If you're sure you can defeat an enemy face to face, toe to toe, a sword approach will do the job.

If you're uncertain whether you have enough resources and reserves to confront an enemy head on, you probably don't, so some sort of alliance of convenience might be in order following a jewel approach.

If you are certain you're not yet powerful enough to confront an enemy, you must bide your time, practice patience, and adapt a mirror approach guerrilla strategy.

MUSASHI CROSSES AT A FORD

"In my doctrine, I dislike preconceived, narrow spirit."
—**Miyamoto Musashi**

People are pretty much the same everywhere. In the East, as in the West, there is much dispute about many things, some arguments earth shaking in their implications, some not worth the spit.

Whether East or West, disagreement is to be found on any street corner, any time of the day or night. This is because, East or West, it's always been easier to disagree than it is to dig up proof.

But in the East—and among the knowledgeable of the West—there is no disagreement as to what are the two greatest treatises on warfare ever written: Sun Tzu's *Ping-Fa* (*The Art of War*) and Miyamoto Musashi's *Go Rin No Sho* (*A Book of Five Rings*).

Musashi is Japan's "Sword Saint," for no greater master of the sword had ever up to his time—or since—mastered the art of the blade as did Musashi. This, too, is not in debate. Love him, hate him, Musashi was—and still is—"da man" when it comes to swordsmanship and/or strategy.

Born in 1584, Musashi slashed his way through some of the most turbulent times Japan has ever seen, when mighty medieval armies of fierce

Samurai slaughtered each other by the tens of thousands up and down the length of the Japanese islands.

During his life Musashi went to war six times and fought in over sixty personal sword fights before the age of twenty-nine, having slain his first opponent at age thirteen.

No one is sure how many to-the-death duels Musashi actually fought during his surprisingly long life, though it is well known he defeated several opponents without ever drawing his sword—using a kitchen ladle, a war-fan, a tree branch, even an oar.

According to one source, Musashi killed over a thousand men in combat and in sword duels (Kosko, 1993:199).

Shortly before the end of his life in 1645, Musashi sat in solitude in a secluded cave writing his strategy masterpiece, *A Book of Five Rings*.

His *A Book of Five Rings* is organized in five sections, named after the traditional Eastern elements: *Earth, Air, Fire, Water,* and *Void*.

According to noted Musashi translator Victor Harris, *A Book of Five Rings* is unique among martial texts in that it deals with "both the strategy of war-fare and the methods of single combat in exactly the same way."

Or, as Musashi himself put it:

> In my strategy, one man is the same as ten thousand, so this strategy is the complete warrior's craft.

But interpretations of Musashi don't stop there. In fact there's no end to the number of Musashi commentators, ranging from the mystical and the philosophical, to how to use Musashi's tactics to get ahead in the financial world.

A Book of Five Rings does indeed lend itself to a variety of interpretations, making it difficult for some, even those with an appreciation of Musashi to find actual modern-day application for his medieval tactics and techniques, let alone try to succinctly sum up the Sword Saint's teachings.

Fortunately, we need only to read carefully to discover that Musashi has done this for us; summing up his philosophy in a succinct (and final) lesson.

He called this "crossing at the ford," a section that appears in the book of *Fire*, and which encapsulates the whole of his strategy for facing life and death in general, and for overcoming enemies in particular—without ever having to draw your sword!

In this one verse, Musashi gives us the key—actually twelve keys—for honing our mind-sword to a fine edge, before then turning that fine focus toward our enemy.

Thus spake Musashi: "I believe this 'crossing at a ford' . . ."

occurs often in a man's lifetime.

Opportunities happen every day. It is human nature that we all too often only see those things we go looking for.

Opportunity knocks all the time, we're usually just too busy bitching or else listening to our inner child crying about how bad life is treating him, how it doesn't recognize what a smart and cute little boy he is, and how nobody will help him change his shitty diaper.

Chance favors the prepared mind. Instead of sitting around bemoaning how "opportunity has passed me by" or how "guys like us never catch a break," we need to prepare and study for the *next time* opportunity knocks on our door.

And, if you should get tired of waiting for Ed McMahon and opportunity to find their way to your Welcome! mat, get up off your fat ass and go stalk him . . . uh, opportunity that is, not Ed McMahon.

It means setting sail even though your friends stay in harbor,

Peer pressure, the need to conform to family and social norms. It's hard enough to screw up your courage, swallow your fear, and gird up your loins when deciding to take a chance by yourself. This difficulty is compounded depending on how many timid and cynical other naysayers you have to convince and carry with you.

All too often we're held back by "toxic relationships," some we choose—our jobs, social and sexual relations—while others are thrust upon us at birth—family, racial, and nationalistic obligations.

Even those who care the most about us are often guilty of holding us back, of telling us how we should just "give up that crazy and dangerous dream" of trying something new before we "end up in the Poor House."

I got news for you, my friend. If this is the kind of advice you're getting from those living under your own roof . . . you're already in a "Poor House"!

It's called *egoism*, and it's not a bad word. It basically means you're not much good to other people until you first get yourself squared away. Often this means rowing out to sea all alone in a little, leaky boat.

Naysayers be damned:

"Nay-sayers seem capable of saying naught but Nay, yet if they could say anything else in any other way they'd eventually make that too a Nay. So, listen not, for Nay is but all the Nay-sayers would say."
—Andre Spearman

knowing the route,

In any undertaking, (1) you first gather intelligence and then (2) you formulate a plan . . . and then (3) you formulate a "plan B."

Plan your route step by step, always leaving enough leeway to allow you to dodge—side to side when need be.

knowing the soundness of your ship

A craftsman is only as good *as* his tool, a warrior only as good as his weapons. Before undertaking any project, trip, whatever, take a realistic accounting of your needs and balance them against the resources you have at your disposal.

Metaphorically, you have to know the soundness of your ship. In other words, you really don't want to set sail in a little leaky rowboat.

And I'm sure you're smart enough to realize "checking the soundness of your ship" includes checking the "soundness" of your crew. The most seaworthy of ships have all too often been scuttled by crews both scurvy and mutinous.

and the favor of the day.

In the *Zendokan Budokai* school of martial arts, the first lesson a student learns is the "Three Knows": know yourself, know your enemy, know your environment.

Having determined to "cross at the ford," you've already realistically assessed your own abilities, side-stepped the naysayers, and double-checked your ship and crew. But you also need to check which way the wind is blowing—literally as well as figuratively.

Whatever innovative idea or bold course of action you have planned, does your immediate environment—physical environment as well as the political environment—support it? Or are your thoughts and actions bound to rub some people the wrong way?

Can the market handle your new product? Indeed, is there a market? And will your new way of looking at things ruffle a few wrong feathers?

What enemies and naysayers will rise against you tomorrow because of the bold steps you take today?

When all the conditions are met,

A final realistic assessment of our recourses, and the resources we've assembled. This is what's called "a reality check" or what combat veterans call a final "weapons check."

and there is perhaps a favorable wind, or a tailwind,

It is always good to have the wind at your back than to try spitting into it. Then again, running against the wind does make you stronger.

A "favorable wind" is also another way of saying "carpe diem"—seize the opportune moment, when wind is fair and confidence high.

Faced with uncertain winds, a good sailor changes his tack. Faced with unfavorable conditions, always be willing to change your tact.

then set sail.

Time to put up or shut up. Change is hard. And there'll always be plenty of people standing around to laugh at you when you fall flat on your face. . . . So don't fall flat on your face.

The taste of not trying is twenty times more bitter in the mouth and gravel in the gut than trying your best and failing could ever be.

It's been said experience is what you get when you didn't get what you wanted. By the same token, excuses are what you *give* when you got nothing because you didn't try everything!

Remember the old Taoist—by now, universal—adage: The journey of a thousand miles begins with a single step.

> *"Step by step walk the thousand-mile road . . . Study strategy over the years and achieve the spirit of the warrior. Today is victory over yourself of yesterday; tomorrow is your victory over lesser men."*
> —**Miyamoto Musashi**

If the wind changes

Clausewitz warned that "No battle plan survives first contact with the enemy."

Ever alert, we adjust, we adapt, and we adopt new ways of thought and action as ephemeral circumstance demands and ultimate victory requires.

within a few miles of your destination,

Most accidents happen in close proximity to our returning home, when we relax prematurely. The game isn't won until the gold is in the bank and your enemy's head is on the silver platter.

The end is important in all things. How many times have we seen the so-called great and famous suddenly brought low in later years by scandal?

On the other hand, look at those who have lived unremarkable lives up to the time of their death but, when Fate demanded it, they died well.

Those are the ones we remember . . . or should. Flight 93.

you must row across the remaining distance

Remember that "plan B" we mentioned? Having committed yourself, your resources, and those who trust in you to lead them to victory—stay the course!

Full effort is full victory. We win by trying.

without sail.

Worst-case scenario: Your friends and family have abandoned you, refusing to take part in your "foolishness." Your resources have been stretched to the limit, perhaps even bankrupt. And the prevailing wind has turned against you or—worse yet—abandoned you to the doldrums. Now you stand alone, wondering what to do. . . .

What *do* you do?

You stand alone.

And you row alone. Read Hemingway's *The Old Man and the Sea*.

And what is our reward for having ventured out, in the face of scorn and storm, both hands on the tiller, both eyes on the far horizon?

We repeat Musashi's promise:

If you attain this spirit, it applies to everyday life. You must always think of crossing at a ford.

Among would-be Samurai in Japan, the martial skills were highly prized but even more prized was the presence of mind. A Samurai in training was constantly subjected to contrived sudden dangers, but if he exercised little cathectic control over his skills and strength he would be severely reproved by his Zen Master."
—Lyman & Scott, 1989:93

Miyamoto Musashi is credited with founding the *Niten-ryu,* the "Two-Swords" school, having mastered the skill of fighting with two full-size Samurai swords at the same time, one in each fist.

Likewise, his "Crossing at a ford" not only advises us on how to live a prudent, productive, and potent life, it also holds clues as to how to prevent our enemies from doing the same.

WHEN "NOH" MEANS "YES"

*"For in and out, above, about, below
'Tis nothing but a Magic Shadow-show,
Play'd in a Box whose candle is the Sun,
Round which we Phantom Figures come and go."*
—*Rubaiyat,* by Omar Khayyam

The Immortal Bard, through his manic-depressive mouthpiece Hamlet, taught us—or is that warned us?—that "All the world's a stage, all the men and women, merely actors."

Always hip-deep in intrigue, the Imperial Japanese court often used small talk about the popular Noh theater to hide what they were really saying—and plotting.

Noh is a quite stylized, even ritualized, form of theater, and Imperial intriguers were all well acquainted with the traditional settings, roles, and plotlines for the most popular Noh plays. How easy it was then to use a reference from a popular Noh drama to scandalize an enemy without doing it

I Believe This "Crossing at a Ford . . ."	*Attack Strategy*
1. *occurs often in a man's lifetime.*	1. Encourage his short-sightedness and his pessimism. Assure him he's right: Opportunity never knocks for guys like him. His best days are behind him. Play him Springsteen's *Glory Days*.
2. *It means setting sail even though your friends stay in harbor,*	2. Encourage his procrastination. Why "set sail" today, the weather might be more favorable tomorrow. His hesitation is your friend.
3. *knowing the route,*	3. He should give in to peer pressure. Convince him he doesn't/ shouldn't/ must not go against tradition and taboo and what his bitching wife is telling him. Better he should "go along to get along". Don't rock the boat.
4. *knowing the soundness of your ship*	4. He needs reliable information—deny him! Restrict his access to true intelligence while fattening him up with heaping helpings of dis-information. He doesn't know (the real) you, he doesn't know himself, and he doesn't know his environment. He is doomed!
5. *and the favor of the day.*	5. Encourage his laxness. Sell him cheap goods and cheaper information. Spike his guns and trip up his messengers. Assure him those cracks in the hull of the ship you are selling him will make it sail faster!
6. *When all the conditions are met,*	6. Convince him his view of the world (no matter how truly screwed up!) is perfect, that he sees things perfectly and it's the rest of the world that's messed up and out of step. Sure, he *should* go against the boss's explicit orders and do it his way . . . the boss will thank him later.
7. *and there is perhaps a favorite wind, or a tailwind,*	7. Assure him it's okay to cut corners, why should he double-check everything, or worry about all the little details? Too much work. Besides, all the "little stuff" usually works itself out in the end.
8. *then set sail.*	8. Agree with him that the wind is favorable when it's not, that black is really white, up is the new down. Conversely, when the wind seems to be at his back, convince him now is not a good time to set sail. (See #2)
9. *If the wind changes*	9. Encourage his stubbornness. Why should he change every time the wind changes. He should stick to his guns . . . like that guy Custer and those men who rode with the Light Brigade.
10. *within a few miles of your destination,*	10. The job's half-finished, convince him it's okay to take a break— you'll handle the rest for him. He's already proven his genius, he deserves a vacation. You'll mind the store. Or, he's already beaten you, you're running away, throwing in the towel. You're no longer a threat to him . . .
11. *you must row across the remaining distance*	11. The wind (and everybody else) has deserted him. He should get out while the gettin's good. Why try to save a sinking ship. He should do the smart thing and abandon ship, cut his losses. It's not his fault, nobody will blame him . . .
12. *without sail.*	12. The wind has deserted him, he has no sail. He was right all along; the world is against him. He was a fool to even try. What an abysmal failure he is. He should do everyone a favor and end it all . . .

face to face. Such references could even be used to alert another person of plottings against him and his house.

Ah! How easily bright small talk can be used to disguise the darkest of plots.

A Western comparison to this would be the difference between your saying, "Oh look at that happy couple over there, just like Romeo and Juliet!"—since everyone understands the reference to the great lovers. However, your friend responds with, "They're more like Delilah and Othello!"

With this allusion to two characters from literature, your friend has warned you that the woman in question is a schemer and not to be trusted (Delilah, betrayer of Samson, from the Bible?), and that her new boyfriend is the jealous type (Shakespeare's Othello, who went mad with jealousy and murdered his wife).

Don't panic. You don't have to be a literature professor, capable of quoting Shakespeare ad nauseam to understand—or participate in—this ploy. We all use allusions, metaphors, and similes in our everyday talk.*

"He is a regular Hercules!" An easy simile. "She looks like she eats at the Karen Carpenter Cafe!" A little more obscure reference.

What if you ask a coworker about the "new guy" and he tells you, "I heard about him from his former associates; they say he's the kinda guy who spends all his off-hours reading Machiavelli." You've just been warned to keep an eye on your job.

"Hey, since you're not dating Sally anymore, think I could ask her out on a date?" Reply, "Sure, if you're the kind of guy who likes living in a soap opera." Translation: she's a drama queen.

"What do think about Bob?" Smirk precedes reply, "Oh, I hear he's a real outdoorsman . . . visits Brokeback Mountain every chance he gets." Hmmm.

> *"Like the Zen ways of tea ceremony, ink drawing, and other arts. Nō suggests the essence of an event or an experience within a carefully delineated art form."* —The New Encyclopedia Britannica, 1986, vol. 28:577

---

*To remember the difference between a metaphor and a simile, repeat this jingle: A metaphor uses "as" and "like" nevermore.

"Noh" (meaning both "talent" and "skill") began in the fourteenth century, having developed from ancient forms of shamanistic dance and drama performed at Shinto shrines and temples. These included *Okina*, an opening dance meant to promote peace and prosperity.

Traveling groups of secular actors soon picked up on the popularity of these plays and eventually took this Noh show on the road. Like their European counterparts, some of these bands of thespians doubled as bands of thieves.

In Japan, Ninja sometimes infiltrated existing bands or else sent their own troupe of players, in order to move about freely through enemy territory, or in order to weasel their way inside a town or a Shogun's castle.

These were similar to the traveling troupe employed by Hamlet to further his revenge against his evil uncle.

Noh (also written Nō) predates Kabuki theater, which wasn't founded until the sixteenth century. Noh also differs from Kabuki in that the former is more subtle in its storytelling. Noh's approach depends more on the use of symbolism and subtle allusions.

Westerners raised on *Rambo* are often disappointed with their first taste of Noh since there seems to be little happening in a Noh drama. This is because Noh is less about the presenting of action than of simile and metaphor.

Noh speaks more to the subconscious. This subtle use of symbols lends itself to attacking your enemy on a subconscious (subliminal) level, under his conscious awareness radar.*

This is why Noh stories are sometimes described as *Yugen* ("dark, obscure"), since for the uninitiated, Noh's intricacy, subtle moral lessons, and overall beauty and symmetry of the story are only partially perceived on a conscious level.

This is why we must cultivate our *"merkura,"* our inner eye that sees beyond outward appearance—both in order to appreciate Noh and, more important, to see through the face-paint and elaborate "play" our enemy is putting on in order to hide his true agenda (his true face!) from us.

Therefore by learning to appreciate the overall staging of *The Five Types of*

---

*For more on how symbols and metaphor have been used down through history to manipulate mankind, read Joseph Campbell's *The Power of Myth* and his *Hero with a Thousand Faces*.

Noh Plays we will more readily be on our guard should an enemy try applying those same settings and situations to his "obscene" advantage.

FYI: The word "obscene" originally meant "off stage."

Moreover, on occasion we ourselves may find it necessary—or merely expedient—to "act obscene," i.e., finding application for our appreciation of Noh, those settings, situations, and subtleties that can "obscenely" benefit us.

Likewise, by mastering *The Six Roles* used in Noh theater we can add yet another useful skill to our already formidable Black Science repertoire, a skill that allows us to, first, instantly recognize when we are being "played" by an enemy and, second, a skill that allows us to take on those roles necessary to help us adapt to any situation and achieve any goal.

And finally, by familiarizing ourselves with the variety of *Masks* used in Noh, we will become better adept at slipping on the "face" du jour that best serves our immediate ends; whether that end be the protection and promotion of self, and/or bring the curtain crashing down onto our enemy's stunned head!

The Five Types of Noh Plays

> *"Freud's unconscious, viewed dramatically, is the backstage of the mind."*
> **—Lyman & Scott, 1989:xi**

Today there are Noh plays known as *Gendai Mono* (lit. "present-day play") that offer various contemporary types of stories, even mixing the traditional five types of Noh plays.

Traditionally, however, Noh plays fall into five recognized types, each with their own particular focus and psychology:

Kami Mono

Kami Mono ("God" and "Spirit" plays) center around the actions of deities (kami) and spirits (rei). Often these deities interact with demon folk, usually *Oni*, mean-spirited animal-skin-wearing brutes, or *Tengu*, half-man half-crow tricksters from whom, not surprising, the Shinobi Ninja trace themselves (see Lung, 1997a).

This kind of Noh scenario could be compared to the spirit and sprite shenanigans going on in Shakespeare's *A Midsummer Night's Dream*, the biblical book of Job, and in Greek epics where the gods take an active interest in (i.e., toy with and manipulate) the fortunes of man.

Shura Mono

Shura Mono (lit. "fighting play"), centers on warriors and conflict.

Katsura Mono

Katsura Mono (lit. "wig play") always have a female antagonist and usually concerns itself with kokoroa ("the human heart"), i.e., love and relationship problems.

Kyojo Mono

Kyojo Mono (lit. "a madwoman play"). The female antagonist has been driven insane from loss, usually the loss of her lover or her child.

Kiri Mono

Also known as *Kichiku Mono* ("final" or "demon" play) has a supernatural theme; death, unsettled ghosts, earthly debts that won't allow the dead to rest in peace.

Each of these types of plays are dominated by a central character whose personality literally sets the stage for the type of play it is. Art imitates life, or is it the other way around?

These five play types easily lent themselves to Imperial Court intriguers identifying these same "personality types" in the real world.

Once this connection was realized, it was a simple matter for them to assign specific characteristics to specific personality types.

Once a person's dominant type was determined, it was—is still!—a simple matter of "flipping the script," disrupting our enemy's choreography and direction. (See pages 181–82.)

Many of us have a tendency to become caught up in—and even subconsciously create for ourselves—situations that reflect our dominant (core) personality type.

For example, type A personalities seek out danger, excitement, and high-risk ventures. So, logically, we'd expect these types of personalities would actually find more of that kind of "action" than a more cautious, less adventuresome, type B personality.

Japanese Name	English	Focus	Central Theme(s)	Shakespeare Comparison
Kami	"Spirit/God"	Gods, spirits	Power, authority	*Richard III* *The Tempest* *Julius Caesar*
Shura Mono	"Fighting play"	Centers on warriors	Contention and conflict	*Henvy V* *Macbeth*
Katsura Mono	"Wig play"	Female antagonist	Relationship issues	*Othello* *Romeo and Juliet* *The Taming of the Shrew* *Much Ado About Nothing*
Kyojo Mono	"Madwoman play"	Loss of lover or child	Loss and Death	*Hamlet* *King Lear*
Kiri/Kichiku	"Final" or "demon"	Supernatural	Reckoning and retribution. Relgiion and guilt	*Merchant of Venice* *The Tempest* *Midsummer Night's Dream*

Figure 20.

Likewise, it isn't uncommon for a man prone to anger to keep finding himself in situations where his anger has more of a likelihood to explode? This is what's called a self-fulfilling prophecy.

All of us play out "scenes" beforehand in our head, mapping out how we'd like a future encounter to go. Some of us are good at it, some not so good.

We do this every time we try to pick up a honey-filled piece of eye candy in a bar; every time we rehearse for a job interview; every time we're trying to work up enough gumption to ask the boss for a raise.

Some people write entire "life scripts" in their heads, the way they feel their life is supposed to play out.

Of course, such people become frustrated, angry at the world—and occasionally homicidal!—when the rest of us fail to recognize their genius and dare to challenge their extravagant expectations.

Name	Personality Focus	Positive Traits	Negative Traits	Approach/Attack
Kami	Authoritarian	Fatherly, benevolent, ambitious, entertaining	Dictatorial, overly ambitious, Will hitch his wagon to any horse to get ahead, *lust*.	Feed his ambitions, cater to his "God Complex." Push him to become even more dictatorial. Encourage him to overextend himself. Lure and corrupt him with promises of power. He can't be trusted, will ally with anyone who promises power.
Shura Mono	Combative	A defender, protector and patriot, straight-forward	Aggressor, fundamentalist *anger*.	Maneuver him into situations where his combative mature will reveal itself. Justify his need to do violence.
Katsura Mono	Relationship-oriented	Believes in love, a giver		
Kyojo Mono	Emotional	Compassio-ate, Empathetic	Unpredictable Unstable *sympathy*.	Suspicious by nature, stir up his jealousy, feed his latent paranoia. Undermine his relationships. Othello.
Kiri/Kichiku	Believer	Faithful, Helpful	Supestitious, *fear*.	Unseen forces move him. Prone to blind faith. Show him a miracle, let him talk to his dear departed mother and he'll follow you anywhere. Steal his lucky rabbit's foot, jinx his ball team. He is haunted by guilt. Prime candidate for blackmail.

Figure 21.

182

I hate to be the one to have to break it to you, but fantasies don't always come true. Furthermore, having unrealistic expectations that the world owes you a living makes you a prime candidate for every "get ahead without sweat," "too good to be true," "you're a member of the master race/God's chosen people" scam that comes your way.

This is why you have to follow both Socrates' and Sun Tzu's number one rule: "Know yourself." Or, in other words, "Check yourself before you wreck yourself!"

Conversely, we want to do everything in our power to encourage our enemy's fantasies. The more fantastic—illogical and unrealistic—the better! (See *Kami*, page 182.)

Our enemy wants to play one role, never suspecting we have another role for him in mind.

He's got Machiavelli tucked snugly and smugly under his arm, thinking he's gonna be playing the role of the prince, when we've already got him outfitted for a frog costume. He's got grandiose plans of playing Hannibal Barca the Conqueror, but, by the time we get done rewriting his script, people are going to be looking at him like he's Hannibal the Cannibal!

Review *"The Art of Seduction"* section in *Mind Control* (Lung, 2006). You remember, where that hot babe you just met wants you to play one role (her future baby's daddy) while you have a completely other role in mind: Johnny-come-lately . . . and often!

The Six Roles

"To beguile the time, look like the time."
—**Shakespeare,** *Macbeth*

Within Noh drama there are six traditional roles, 3 principle roles and 3 peripheral or subsidiary. The three main roles are (1) *Shite*, principle actors around whom the action centers; (2) *Waki,* secondary and subordinate actors; and (3) *Kyogen*, who acts as the narrator. Kyogen can also refer to those actors who perform humorous sketches in between short plays.

Then come the three subsidiary roles: (4)*Tsure*, attendant(s) who interact to support the main players; (5) *Kokate*, "boy," often used as a go-between, or through whose "naive" eyes we view the unfolding drama; and (6) *Tomo*, "walk-ons," nonspeaking parts.

In the real world, "Tomo" are those "little people" we talked about in

Mind Control, those worker bees in the background we seldom notice but who—either because we've somehow slighted them or because our overly generous enemies have "turned" them—can all too easily be used against us; from the pharmacist's assistant who tells your enemy what you're allergic to, to the parking valet who—for a price—hides that packet of heroin or that GPS tracking device in your glove compartment.

Japanese Imperial Court intriguers knew that these stage roles could all too easily bleed over into, and be applied to, life outside the theater. They also became adept at first recognizing the role a (targeted) person had chosen for themselves, and then "recasting" that person into the roles that better suited the intriguer's script. (See figure 22.)

In real life we all have our different roles to play, and not just the six allotted us by Noh. Roles vary from time to time, place to place but, consistently, there are three ways we come by these roles:

First, we take on roles *consciously*, adopting and adapting those roles that best fit our goal(s).

Second, we also take on roles *unconsciously by ourselves*, because it's our nature, and we're inexorably drawn to certain roles, even though we are unaware of our unconscious reasons for choosing such roles.

Finally, we assume roles *unconsciously, but from others*, when a role is "thrust" upon us by others. In turn, the person foisting this role upon us might be doing so deliberately, or else may be unconscious of their actions themselves.

For example, a father dies, a young, inexperienced child is forced to assume the role and responsibilities of the "man of the house." Or in relationships, where you (unconsciously) turn your new bride into your surrogate mother, and then you're flabbergasted when she takes to the role like a duck to water and starts treating you like her mother treated her father! Turn about being fair play.

The Masks We Wear

> *"False face must hide what false heart doth know."*
> —Shakespeare, *Macbeth*

The key to successfully pulling off a flawless performance hinges on not only acting like the character you're portraying, but looking like that charac-

ROLE SHIFTING

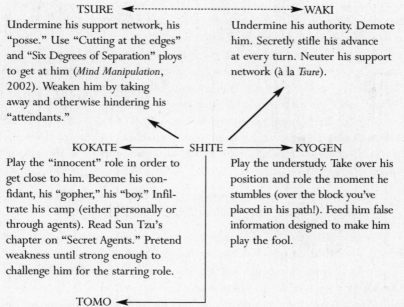

TSURE ◄------------------------------------►WAKI

Undermine his support network, his "posse." Use "Cutting at the edges" and "Six Degrees of Separation" ploys to get at him (*Mind Manipulation*, 2002). Weaken him by taking away and otherwise hindering his "attendants."

Undermine his authority. Demote him. Secretly stifle his advance at every turn. Neuter his support network (à la *Tsure*).

KOKATE ◄——— SHITE ———► KYOGEN

Play the "innocent" role in order to get close to him. Become his confidant, his "gopher," his "boy." Infiltrate his camp (either personally or through agents). Read Sun Tzu's chapter on "Secret Agents." Pretend weakness until strong enough to challenge him for the starring role.

Play the understudy. Take over his position and role the moment he stumbles (over the block you've placed in his path!). Feed him false information designed to make him play the fool.

TOMO ◄———

Make friends with the "walk-ons" in his life, the "little people" he interacts with every day but hardly notices. Pump them for information, inflame any petty grievances they have against him for real or imagined slights. Where such actual slights do not already exist, create them.

Figure 22.

ter as well. This requires the correct dress and deportment and, most importantly, wearing the right false face. (See "The Art of Disguise" in *Nine Halls of Death,* Citadel, 2007.)

Noh actors accomplish this through the use of masks—friendly and fearsome.

There are 125 named varieties of Noh masks: males, females, elderly characters, gods, goddesses, devils, and goblins.

Noh masks are made of wood, coated with plaster, and then lacquered and gilded. Even the color of the masks holds significance to a knowledgeable audience. White means a corrupt ruler. Red, a righteous man. Black is always reserved for violent and brutal villains.

Japanese audiences recognized these masks and understood the underlying attitude, emotions, and motivations inherent in each traditional mask. The same way a Black Science graduate can tell by a person's expressions, "shadow-talk" and "shadow-walk" what they are thinking.

In their masterpiece of sociological insight, *A Sociology Of the Absurd* (1989), Stanford Lyman and Marvin Scott figuratively and literally pull the false face off society by revealing how we all, in one way or another, wear "masks":

> In every social encounter a person brings a "face" or "mask"—
> which constitutes the value he claims for himself. (p. 90)

Far from being a universal negative, wearing the appropriate mask in the appropriate season seems to be the only glue holding civilization together.

Masks allow us to establish identity in an increasingly politically correct "don't ask, don't tell" world where you're free to pretend to be anything— and anybody—you want as long as you don't blow the cover of anybody else busy being anything—and anybody—they want to be!

Unfortunately, once we take on a specific role, whether suited to that role or not, we can all too easily become "caught up" in that identity/role.

Having put considerable emotional investment into the role, we then become trapped, finding ourselves spending more time "maintaining" that role than in enjoying the fruits of "being" that role.

This is what Lyman and Scott call "characterological survival" i.e., keeping up your front (p. 83).

This is especially true when someone (an enemy, a rival) decides to challenge us for the role we've chosen.

It isn't complicated: Walk around with your chest all puffed up, pretending to be (i.e., playing the role of) a tough guy, and, sooner or later, some genuinely tough guy is gonna come along and "challenge" your claim to the role you've chosen.

Or, a young—perhaps inexperienced—girl decides she's going to play the role of "sexy vamp," complete with false face paint and a slinky dress cut all the way up to "thar"! What are the chances of more than one barfly calling her bluff, "challenging" her to "put up or shut up"?

Beware lest you become the mask.

In other words, there's a danger we'll get caught up in our playacting and start believing our own line of bull. According to Lyman and Scott:

> For the rebel, art does not imitate life. It becomes life. The rebel's very existence is theater. And when the rebel begins to regard clothes as costumes, facilities of all kinds as props, and streets as stages, he is capable of wreaking havoc in the social world—as his dramaturgical innovations break down the line between theater and taken-for-granted reality. But in the process of destroying the world, he has made himself. (p. 196)

An even worse scenario: We get caught up in someone else's playacting, a charismatic con man, a crooked politician (yes, I realize that's redundant), or a crazed cult leader.

The only thing worse than forgetting you're wearing a costume is to actually start taking the other guy's costume too seriously.

Remember from *Mind Control* that any "power" that comes from outside yourself (like the uniform you wear), or else power given you by another, isn't real power.

The story is oft told of accomplished cross-dressing actor/actress RuPaul once being approached by a leathered-out, heavily tattooed outlaw biker-type who felt the need to question RuPaul's dressing in . . . well, dresses.

The ever-composed entertainer looked the biker up and down—fudging his fashion sense, or lack thereof—before explaining, "It's *all* drag, honey."

In other words, we all dress for effect, to portray the character role we want the world to see. And, of course, part of that is choosing the right outfit (or ensemble, if you will) to fit the time, place, and effect we are trying to project. Whether that time and place be a drag queen ball or a barroom full of bikers, we outfit ourselves for success . . . and survival!

In fact, the word "outfit" originally comes from "outside fit" i.e., the clothes (and masks) we wear for the world.

Your "drag" or "outfit," depending on your preference (heh-heh-heh), can range from the coveralls you wear and the bucket you push that allows you to mop your way past the guard in the bank lobby without arousing suspicion, to the three-piece suit and false FBI badge you flash to convince that old lady into withdrawing her life savings so you can use it to bait a trap to catch some scumbag whose stooping so low as to rip off little old ladies of their life savings.

Guerrilla Theater

Sitting in a darkened theater, we can all too easily get caught up in the action on stage or on that silver screen up there to the point to where sometimes it's hard to tell where the play ends and the reality begins.

Life imitates art and sometimes events and encounters happen right in front of our eyes and we become caught up in them, never suspecting (until too late!) that these events are being staged for our benefit.

In the 1960s, counterculture trickster Abbie Hoffman and his "Yippies" perfected "guerrilla theater."

Imagine you're standing at a bus stop when suddenly a heated argument about the Vietnam War breaks out between two people standing nearby. You'd just have to listen, huh? Probably tell your friends around the water cooler all about it, huh? Well that's exactly what Abbie and his Yippie comrade *staging* the argument wanted to accomplish.

Abbie and his friends staged every manner of "plays" just like this—from impromtu two-person bus stop skits, to major productions involving hundreds of "actors," like the disruption of the 1968 Democratic National Convention and accompanying street protests—followed soon after by the infamous "Chicago Seven" trial—still more "guerrilla theater"!

This tactic has recently been resurrected for those slick, deliberately disturbing "*Truth/Crazy* World" anti-tobacco ads where unsuspecting pedestrians are invited into a "Crazy World Carnival" where all the attractions turn out to be thinly disguised stabs at the tobacco industry's hypocrisy.

Good for the goose, good for the gander. It recently came to light that companies like PayPerPost.com and MindComet.com (a Tampa-based interactive ad agency) had begun paying bloggers to write nice things about corporate sponsors and "talk up" specific products on their individual blog sites. Most of these bloggers never let on to their readers that they're actually "working" for the product's PR firm. ("Polluting the Blogosphere" by Jon Fine. *Business Week*, July 10, 2006:20.)

This kind of word-of-mouth ploy has been used by advertisers ever since there's been something to advertise. The Internet just has a bigger mouth.

Getting people to talk about a product is the first step in getting them to buy that product—no matter if that product is a new soft drink or your cult's hard-sell spiel.

Confidence men are masters at "setting the stage," catching us up in a scenario where everyone involved is in on the scam, except "the mark." Watch Robert Redford and Paul Newman's *The Sting* (1973). Watch it twice . . . and take notes.

Medieval Ninja were masters when it came to setting the stage, preparing area and "actors" for future operations. For example, a group of travelers on the way to the Shogun's castle are attacked by masked highwaymen but are saved at the last minute by a brave young ronin. To insure their safe passage, the young ronin agrees to see them safely to the castle.

Predictably, once the Shogun hears from the travelers how this young ronin saved then, on the spot, the Shogun enlists the ronin to work for him.

It doesn't take a genius to figure out that the local Ninja clan has just succeeded in infiltrating one of their agents inside the Shogun's castle and, more important, into his confidence.

Police use similar ploys on dumb criminals, staging elaborate comings and goings: stenographers who seem to be taking a "confession" from your partner-in-crime in the next room over; your partner getting special drink and food, perhaps extra phone calls?

Then there's the old "good cop / bad cop" setup, all clever staging and dialogue designed to confuse criminals into falling over each other to be the first to "cut a deal."*

We study the various types of "plays" (from the theater to the street), roles assumed and masks presumed, first in order to use them effectively and, second, to keep our enemy from using them effectively.

The Show Must Go On

In any staged production there are the actors we see, then there are the scores of support personnel behind the Black Curtain we don't see, but without whom the show could not go on.

When we can't cast ourselves in the main role, or even get out on stage, there are still five things we can do:

---

*For more on the Art of Interrogation, see my books *Black Science* (Paladin Press, 2001) and *Mind Manipulation* (Citadel, 2002).

First, *play the prop-master*, using props (tools, disguises, and costumes) to our advantage. People "respect" (i.e., fear) a uniform; this can work to our advantage since so many people will automatically obey a uniform (police, fireman, security guard, military), or someone with impressive looking credentials (badges, bills of sale, college degrees) without question.

People can be props, too. That pretty-boy actor, the one who the tabloids keep wondering out loud why he's never married, or ever even been seen with a girlfriend for that matter? His agent makes darn sure he shows up on the red carpet at every awards show with a buxom blond bombshell dripping off his arm.

Accused of neglecting his minority constituents, the next time you see that wily white politician you can bet he'll be pumpin' the flesh and/or receiving an honorary humanitarian award down at that NAACP banquet-slash-photo op.

So we use props to our advantage while at the same time sabotaging our enemies props. Can you say "wardrobe malfunction"?

Scenario: The representative from a rival company is on his way to a meeting where both you and he will make presentations for a new product line. But only one of you can win the contract. How sad that his car breaks down on the freeway and he misses the meeting. Or what if his secretary has a breakdown, vehicle-wise or, better yet, mental health–wise, and she can't make it to the meeting? Won't this improve your chances of taking center stage?

Second, *play the critic* by refusing to acknowledge an enemy's acting ability, exposing him for the poseur he is.

We draw attention to the flaws and faux pas in his acting technique. We openly challenge his facts and statistics; we put him on the defensive and keep him there. As long as an enemy is kept on the defensive, he can't rally his forces to go on the offensive.

Third, *play the agent*. An alternate script calls for us to use the Judo principle: instead of openly opposing him, feed his fantasy, become his confidant.

He wants to "play" leader, give him what he wants. Feed his delusions, encourage him to overextend himself. Convince him he can handle it. Give him enough rope . . . a *"Kami"* and *"Shura Mono"* ploy (see page 182).

Fourth, *play the director*. Feed him misquote and miscue, encourage his mistakes. Make him miss his mark and stumble over his lines. As soon as he memorizes his lines, rewrite the script. Break the routine to break his concentration. Keep him doubtful and in the dark. The "Mushroom Treatment."*

Fifth, *play the understudy*. Whatever theater you play in in life, there's always some eager young understudy waiting in the wings—salivating—just praying you'll stumble over your lines and literally break a leg.

As soon as . . . uh, I mean, *if* something should ever happen to the "star," some highly improbable accident, e.g., he should succumb to suspiciously good health, whatever, you will be there to step into his shoes, to save the day for the rest of the cast. After all, the show must go on.

Duh! Can you say "identity theft"!

> *"He wears a mask, and his face grows to fit it."*
> —George Orwell

"8-9-3"

> *"You stand in the way not merely of an individual, but of a mighty organization, the full extent of which you, with all your cleverness, have been unable to realize. You must stand clear . . . or be trodden underfoot."*
> —Professor Moriarty

Yakuza. The Japanese "Mafia." Some see them as a benevolent "Freemason"-like organization, some Yakuza groups even openly advertise themselves as such. Some see them as shugo-rei "guardian angels," Robin Hoods who flout the law and help the needy.

Others know them not as saving angels but as Oni-demons! Or, at the very least, *kanjin* ("bad guys") with a *kuro-kakure,* a dark and hidden agenda.

Yakuza clans began appearing after Japan's *Sengoku,* "Warring States" era (1467–1572), when the Tokugawa Shogun finally succeeded in uniting Japan under one iron fist, ending years of internecine warfare between rival Samurai, Ninja clans, religious factions, and just about anybody else with a sharp stick.

---

*The Mushroom Treatment, keep 'em in the dark and feed 'em plenty of bullshit!"

Some of these Yakuza clans were founded by Samurai, now ronin, who had been defeated and dispossessed by the Tokugawa regime. Some Yakuza clans were formed by out-of-work Ninja after their lucrative mercenary work dried up as the Tokugawa sun rose. And some Yakuza clans matured out of already established pseudo-criminal pastimes such as gambling and prostitution.

The name *"Yakuza"* comes from the unlucky numbers "8-9-3," the worst score possible in a popular underworld game called *hanafuda* ("flower cards"). This is analogous to Western criminals tattooing "Born to Lose" on their arms. The typical criminal mind's attempt to use "reverse psychology," i.e., wallowing in the shitty hand life has dealt him—as opposed to getting a real job!

Today it's estimated there are tens of thousands of Yakuza members in Japan, and looking out for Yakuza "interests" overseas.

Yakuza groups are called *uji* (lit. "group"). Its members, *ujiko,* are led by a *uji-no-kami,* a chief of a particular uji.*

Groups of uji make up recognized—and feared!—Yakuza families and clans.

A Yakuza family is lorded over by a Kuromaku, a "Black Curtain." This word "Kuromaku" originated in classic Kabuki theater where an unseen wire-puller controls the stage by manipulating props and players from behind a black curtain.

Today "Black Curtain" likewise connotes a powerful godfather, or fixer who operates behind the scenes (Kaplan & Dubro, 1986:78).

Today, it is said not a grain of rice falls to the floor in Japan that there is not at least two Yakuza uji there to fight over it. This is a nice way of saying (warning us?) that Yakuza have their hands in every pocket, purse, and potential moneymaking venture conceivable—legal or otherwise.

There are Turukuos who run "harmless" houses of prostitution. And then there are sokaiya, gangsters who specialize in shaking down Japanese corporations. For example, a Yakuza gang known as the "Man with 21 Faces" once extorted six major food companies by threatening to lace their products with cyanide. In fact, some of these products, already on store shelves, were found to have "warning labels" placed on them by the extortionists: "Danger. Contains poison. Eat this and die" (Kaplan & Dubro, p. 180).

---

*In this case "kami" means "superior," not "deity," but still implies god-like power.

At one time or another various Yakuza clans have been caught between the sheets with any number of shadowy conspiracies, cliques, and cults: from the infamous *Kokuryu-kai* (Black Dragon Society), to working for the Tokko (National "Thought Police") during World War II, to allying themselves with cults like Sun Moon's Unification Church—called *Genri Undo* in Japan, The Unification Church has since been linked to the Korean CIA.

Since many Yakuza uji trace themselves (factually or fancifully) back to Samurai warriors, it's not surprising to find them embracing both the "silk" and the "steel" aspects of Samurai philosophy.

To outsiders, Yakuza might be *gutentai,* "hoodlums," but within their own brotherhood they have their own unforgiving code of honor, similar to the Code Bushido of the Samurai. This Yakuza code includes:

- *Giri:* fulfilling all duty and obligations to self and clan;*
- *Shojiki:* honesty, veracity, and frankness, at least among fellow Yakuza brothers; and
- *Shojin:* devotion to the uji, diligence in carrying out instructions from a superior, and performing acts of restitution and purification for mistakes. The latter ranging from the ritual cutting off of a finger, to performing an assassination.†

The worst thing for a Yakuza soldier is to be declared dasoku, a word that literally means "a snake's legs," i.e., something totally useless. Better dead than useless.

Despite their fearsome reputations, the Yakuza ideal is to maintain "balance" and promote "harmony." Thus a "balance" of "silk" and "steel" methods are used to accomplish their goals—restoring balance, maintaining harmony. Bloodshed is bad for business.‡

This *"Silk and Steel"* approach applies even when it comes to the Yakuza's four main areas of operation: (1)*Tobaku* (gambling), (2) *Baishun Torimochi* (pimping), (3) *Kyohaku* (extortion), and *Satsujin* (homicide).

---

*Duty is what you owe to yourself. Obligation is what others try to foist on you.

†See Robert Mitchum's *Yakuza* (1975) and the 1989 movie *Black Rain* with Michael Douglas.

‡Unless the blood is spilled in so spectacular a way as to send a message to others to modify their behavior, i.e., see things *your* way! (See *Theatre of Hell: Dr. Lung's Complete Guide to Torture*, Loompanics Unlimited, 2003.)

The Japanese in general, and the Yakuza in particular, see these first two, gambling and prostitution, as harmless vices (albeit, not to be talked about in polite company).

So far as extortion is concerned, Yakuza view this as, first, a business opportunity (i.e., my enemy's weakness and indiscretion has added to my strength by giving me an opening through which to stab at him). Second, extorting someone who has done something dishonorable is seen as karma and even as setting an example for others to be more discreet (or, at the very least, destroy the videotape afterward!).

And the fourth, murder? That's also seen as the price of doing business, and as a way of restoring balance, of righting a wrong (an insult, most likely) and of exacting "an eye-for-an-eye."

And, as already mentioned, one of your enemy's dying a most horrible death sends a point-blank message to your other enemies.

On a more cerebral "less mess to clean up" level, there are psychological ploys to be derived from these four.

Black Science graduates will have already spotted correspondences between these four methods of "doing business" and "The Killer B's" (from *Mind Control*), as well as the Five Warning F.L.A.G.S. (see figure 23).

THE BLACK MIST

"The middle ground was, by design, inscrutable. One's deeper motivations remained concealed. One could play the spy game almost as a kind of private joke."
—Dick Russell, *The Man Who Knew Too Much*, 1992:143

The theme of the inter-connectedness of all things runs through the whole of Asian thought. We find this philosophy in Indian *Auyurveda* (see "Indian Insights," *Mind Control,* 2006:197), in Chinese Taoism, Ibid., p. 62), with the Black Crows of Vietnamese Cao Dai (Lung, 2002), and throughout the history of Buddhism.

But nowhere is this inter-connectedness more evident than in the hidden collusion and conspiracies linking shadowy Far Eastern cliques and cadre.

"YAKUZA STRATEGY"

Enterprise	Personality	Warning F.L.A.G.S.	"Killer B's"	Strategies
Tobaku (Gambling)	Risk taker	Greed	Brainwash, Bribery	Tempt him with gain, something for nothing. Convince him to take a chance, bet it all on a single roll of the dice.
Baishun Turimochi (Pimping)	Impulsive	Lust	Blind (with seduction and lust)	Cater to his appetites (e.g., sex and drugs), then catch him with his pants down. Samson. Entice and entrap.
Kyohaku (Extortion)	Cautious (Suspicious)	Fear and Sympathy	Bully, Bloodties, Blackmail	Make him indebted to you. Remind him of his obligations and the secrets you know about him (bloodties).
Satsujin (Murder, also means "Insight")	Stubborn	Anger	Bury	Encourage him to stand his ground even in the face of impossible odds (i.e., help him dig his own grave). Trap him in a situation where his anger (and refusal to compromise) spells his doom.

Figure 23.

In Japan they call this *Kuroi Kiri*, the "Black Mist," a dark record of all the dirty tricks, corruption, and organized crime in Japan.

But guilt for this foul miasma of mind manipulation, malice aforethought, and outright murder overshadowing Asia as a whole cannot be Japan's burden alone.

The Asian "old school" of intrigue and intelligence, double-dealing and poisoned dirks in the dark, stretches far back from modern times, back through the Middle Ages, all the way to ancient times, and includes as its alumni those adept at Chinese counter-espionage, Indian intrigue, and Tibetan treachery.

However, nowadays there seems to be less paranoia about such old school adepts (surely they've all died out by now? heh-heh-heh) and more concern about New Age Far Eastern cults and philosophies, all of whom seem to serve some mysterious "New World Order" bent on sinister domination—today individual minds, tomorrow the world!

But is there really anything to fear from the inroads Far Eastern philosophies have made/are making into the West?

There's no argument cults like Bhagwan Shree Rajneesh and Moon's Unification Church have given Eastern spiritualism a black eye. But for every opportunistic Asian cult leader, there are a dozen sincere guru, sifu, and sensei eager to share their Zen experiences with us, more than willing to teach us their martial arts.

So where is the connection between the "inscrutable" Fu Manchu old school masterminds of yesteryear and the—possibly even more sinister—pimps and pushers of that mind-raping New Age philosophy suspected of kowtowing daily to the more dreaded bugaboo of the New World Order?

Who's to say for certain? Right. Dr. Lung.

Ancient myth mixes freely with modern media ratings to distort any surviving vestige of any shred of information remotely resembling "the truth"; so much of that distortion being deliberate disinformation spread by the very groups under scrutiny—sowing confusion in the minds of inspectors, superstitious fear in the hearts of their enemies.

Would it interest you to know that one man, a Japanese professor, is credited with—or is that "accused of?"—coining both terms *"New Age"* and *"New World Order"* . . . while, at the same time, mixing ancient Asian mind-control techniques with twentieth-century technology?

The Pacific Theater

Shortly after Pearl Harbor, George Estabrooks, then chairman of the Department of Psychology at New York's Colgate University, was called to Washington, D.C., by the War Department who wanted to know how Germany and Japan might use hypnosis and other sorts of mind manipulation as weapons of war.

Although the term "brainwashing" would not be coined until several years later, the OSS (forerunner of the CIA) and U.S. Military Intelligence were already hot on the trail of such techniques, for defensive as well as potential offensive use.*

From the moment of its inception in 1947, the CIA was already deeply invested in a variety of wide-ranging mind control programs, all of which would eventually come under the umbrella designation "MK-ULTRA."

After the war, Estabrooks published his 1948 best seller *Hypnosis*. In it he discusses ways to hypnotize agents (remember, the word "brainwashing" hadn't been invented yet) to the point where they would not only be immune to interrogation, up to and including torture, but also capable (while under post-hypnotic suggestion) of committing treason and even murder.

Word has it, Estabrooks had perfected a method by which multiple personalities could be both caused and cured by hypnosis (Russell, 1992:387).

More on this technique in a minute.

The West came out of World War II with a new respect (fear!) for the role intelligence played in winning battles. The hunt was now on in earnest to improve intelligence gathering: You needed to field more and better agents to counter your enemy's more and better agents, as well as find more and better ways to manipulate and ultimately control your enemy's mind.

It's also no secret the United States came out of World War II realizing, to paraphrase Patton: "We killed the wrong goddamn beast!" In other words, the West had successfully beaten down the black beast of Fascism, at the high cost of further feeding the ravenous red beast of Communism.

In order to counter growing Communist influence in the conquered

---

*Contrary to popular lore, the term "brainwashing" was not invented to explain American POWs choosing to remain in North Korea at the end of the conflict. Edward Hunter, ex–OSS/CIA World War II propaganda specialist turned author, coined the term brainwashing in a September 1950 article appearing in *The Miami News* titled "Brain-washing: Tactics force Chinese into ranks of Communist Party."

Axis nations, in its haste to put the defeated Axis nations back together again, the West cut corners. They also cut loose a lot of bad guys who maybe shouldn't have been cut loose at all.

In Germany, America co-opted Nazi spy networks intact, "forgiving" indiscretions during the war in favor of fighting Communism. The most infamous being the Gelan spy network, originally set up by Reinhard Gehlen, former chief of Nazi Intelligence.

In Sicily, the U.S. Military smoked more than a few fine Cubans with Lucky Luciano and his Mafia.

After World War II, the United States held the big broom, so why should we be surprised so many war crimes were swept under the rug?

We used German scientists to help us build better A-bombs. And we also used—and were used in turn—by Asian experts who knew how to build better "mind-bombs."

Asia has always been fertile ground for spying. From Chinese *Lin Kuei* and *Moshuh Nanren*, to Rudyard Kipling and the *Hircarrah* of India, to Tibetan *sDop sDop*, to Sidney Riley, the *Ace of Spies*.

Sidney Riley has been dubbed the "Ace of Spies" for good reason. The template for James Bond and Simon Templar, it is from this real-life spy that we get the phrase "living the life of Riley."

Russian by birth (or was he a Polish Jew?), Riley worked for the fledgling British Secret Service (or was it for the Germans?).

What is known for certain is that, during the Russo-Japanese War (1904–1905) Riley gave secret plans for the Russian fleet moored at Port Arthur to the Japanese, allowing the latter to sink the former in a daring surprise attack, so successful the Japanese would use the same tactic forty years later at Pearl Harbor.

This "get a dog to eat a dog" ploy was done with the blessing of British Intelligence. The fulfillment of the old axiom: "The enemy of my enemy is my friend." Thus the British got the Japanese (a potential enemy) to destroy the fleet of Britain's only real sea rival in Europe—Russia.

In the 1920s Sidney Riley worked for, and by some accounts helped found, a secret organization called the Trust, helping smuggle expatriated white Russians back into Russia, ostensibly to build a resistance movement against the Soviets.

Only years later was it discovered that the Trust was actually an elaborate trap set up by Black Science mastermind and Soviet security chief Felix Dzerzhinsky. Those returning were never heard from again.

Riley himself reportedly fell victim to this ploy, having been lured back into Soviet Russia by the Trust. Although . . . another version of this story alleges Riley was a Soviet agent all along and that, rather than having been executed, he merely retired back into the loving arms of Mother Russia.

Following in Riley's Asian footsteps, there's the infamous Dr. Richard Sorge, a German who, between 1929 and 1941, was the leader of a notorious Soviet spy ring operating in both China and Japan. Like Riley, Sorge was reportedly exposed and executed in 1944. But, like Riley, you can never really be sure with these spy types . . .

Not surprising, interest in mind-manipulation tactics and techniques soared in the West when it was realized the Cold War was shaping up to be a war of nerves.

Just as they'd successfully done in Germany and Italy, in post-war Japan the OSS (and later the CIA) signed up not just Yakuza Mafia types, but also former members of the feared Kempai Tai (Japan's version of the Gestapo). U.S. Intelligence also employed former administrators and agents of Japan's infamous "Thought Control Police"—all in the name of fighting Communism (Kaplan & Dubrow, 1986):

> Thought Control agents, purged and purged again, keep reappearing in positions of responsibility—often with American encouragement. (p. 57)

It's also not surprising that U.S. Intelligence "looked the other way" in their rush to obtain any edge against Soviet Intelligence.

Just as they'd brought over German defectors and even former Nazi scientists to help work on the A-bomb to help us win the arms race, so too U.S. Intelligence co-opted Asian mind-control experts to help us in the "mind war" (Russell, 1992):

> In the mid-1950s, the CIA-funded MIT Center for International Studies . . . brought over a group from Japan to embark upon a major study of the country's "Science of Thought." It examined how Japan's once notorious "Thought Police," in a

prewar form of brainwashing, had broken down the identities of communists and brought them around to the imperial viewpoints. (p. 145)

Former Japanese Military Intelligence officers, scientists, and sundry scam artists, all eagerly signed on board the good ship U.S.S. *Forgive-n-Forget.* All had their reasons. A few actually did hate Commies. Others were just eager to avoid attending Tojo's necktie party. More than a few dreamed of getting their Samurai groove back.

Others had deeper—darker—agenda. For such men, the dust would never fully settle over Hiroshima and Nagasaki. But they were smart enough to learn the lyrics to "Yankee Doodle Dandy," learn to tell the difference between a squeeze-play and a sacrifice-fly and, most important, they were smart enough to recognize the advantage of playing one enemy off against another.

In Japan, such men are known as Zetsujin.

The Zetsujin

"He is the Napoleon of crime. . . . He is the organizer of half that is evil and nearly all that is undetected in this great city. He is a genius, a philosopher, and abstract thinker. He has a brain of the first order. He sits motionless, like a spider in the center of its web, but that web has a thousand radiations, and he knows well every quiver of each of them. He does little himself. He only plans. But his agents are numerous and splendidly organized. Is there a crime to be done, a paper to be abstracted, we will say, a house to be rifled, a man to be removed—the word is passed to the Professor, the matter is organized and carried out. The agent may be caught. In that case money is found for his bail or his defense. But the central power which uses the agent is never caught—never so much as suspected."

—Sir Arthur Conan Doyle, *Memoirs of Sherlock Holmes,* **1893**

In Japanese, the word *"zetsujin"* means the "offspring of a talkative tongue," for example, debates and arguments, and the words used to accomplish both. Zetsujin is sometimes used to mean the tongue itself.

An accomplished smooth-talker(a BS artist!) is thus a Zetsujin.

For Zetsujin, words are truly dangerous weapons (see "Word Weapons," *Mind Control,* 2006:110).

Tokushiro Hattori was Zetsujin. As the private secretary to Japanese warlord Tojo, Hattori was a leading figure in wartime intelligence, running a special intelligence operation dubbed *the "Hattori Group."*

Like the Gehlen Nazi spy network in Germany, after the war in the Pacific, the Hattori Group was recruited lock, stock, and silenced gun-barrels to work for U.S. Intelligence.

FYI: During Japan's middle ages, the Hattori Clan was one of the more infamous of the Shinobi Ninja clans; hiring their services out to the highest bidder, often working for rival Samurai camps at the same time, playing one off against the other. Smart strategy since, the more time Samurai spent cutting up each other, the less time they had to unite against the Shinobi.

Legend has it the Grandmaster of the Hattori Clan hated the Grandmaster of a nearby rival Ninja clan, with Ninja from the two clans always bumping heads. For example, if the rival clan took a contract to kill a particular daimyo, Hattori's clan would hire themselves out as bodyguards protecting that targeted Samurai lord.

Only after Grandmaster Hattori's death (of old age) was it discovered he'd actually been the Grandmaster of *both* Ninja clans; having created two completely separate identities replete with two separate families. The "enmity" between the two clans was merely a ruse to "stir up business!" (See Lung, 1997a; and Lung & Tucker, 2007.)

According to one source, the modern Hattori was just as wily as his medieval namesake.

Even while working for U.S. Intelligence, supposedly helping fight Communism, Hattori's real agenda in life was to exploit the American-Soviet conflict to Japan's advantage (cf. *Hattori: Sheathing the Sword,* by Meirion and Susie Harries. Macmillian, 1987:226).

But Tokushiro Hattori was by far not the only Zetsujin smooth-talker and shapeshifter to pledge allegiance to American baseball while, all the while, playing ball with the Russians. Nor was he even the best at it.

Prior to World War II, a certain professor Chikao Fujisawa had already made a reputation for himself as one of Japan's leading intellectuals. This distinction led him to serve as Japan's representative to the League of

Nations (failed forerunner of today's United Nations, an organization some see as synonymous with the New World Order—a phrase Fujisawa is credited with coining in 1942).

Fujisawa was a staunchly Right-Wing Nationalist, Japan's version of a Fascist, and appears to have had links to—if not actually having been a member of—the infamous Black Dragon Society (see Lung, 2002a).

During the war, Fujisawa was chief of the research department of the Imperial Rule Assistance Association (Japan's only approved political party). His job was overseeing propaganda aimed at foreign countries in general, and at American and British prisoners in particular.

In 1942, Fujisawa committed his major theories to writing. In English his treatise was titled *On the Divine Mission of Nippon: A Prophecy of the Dawn of a New Age*. In it, Fujisawa coins the terms "New Age" and "New World Order."

Fujisawa's theories: Japan was the original motherland of all human beings and was destined to save the world by uniting all nations under the benevolent rule of the Japanese emperor. The Imperial Gods of Japan had commanded this.

Fujisawa dismissed capitalism and Communism but had nothing but praise for Nazism and Fascism. The latter he believed had much in common with the Japanese principle of "Musubi" ("one-in-many").*

Fujisawa was convinced the Axis powers of Japan, Italy, and Germany were a New World Order meant to sweep aside the old world order. He also believed Hitler had been influenced by Confucius (Russell p. 135).

With bigger fish to fry, an old carp like Fujisawa hardly merited a second look from the victorious Americans. So the professor went back to his writings and to teaching.

Before long he'd quietly started a private study group of students—protégé, if you will—who hung on every word coming out of his mouth.

Eventually the "professor" widened his circle of admiring students to include not just impressionable young Japanese minds, but foreigners as well; wide-eyed students from the Soviet Union and other Communist bloc nations sitting side-by-side with Western students.

As we will see in a minute, one of those promising students included an American named *Lee Harvey Oswald*.

---

*"Musubi-economy" has come to dominate today's world (Russell, 1992:142).

Fujisawa's "harmless" study group has been compared to the infamous medieval "Grand Lodge at Cairo," the ostensibly Islamic university that produced movers and shakers as diverse as Omar Khayyam, sensitive poet and Hasan ibn Sabbah—sinister "Old Man of the Mountain"—founder of the dreaded Assassin cult.

For more on the mysterious Grand Lodge at Cairo and its influential alumni—the mysterious as well as the murderous—see *Assassin! Secrets of the Cult of the Assassins* by Dr. Haha Lung, Paladin Press, 1997.

So who, besides himself, was Professor Fujisawa working for after the war? Well, for one, Fujisawa was reportedly a member of a secretive intelligence agency known as *Hai/Wai*, described by one source as "an unknown Oriental intelligence service" (Russell 1992:166).

There was speculation that Fujisawa had sources within the U.S. Intelligence community feeding him information, even as he was (supposed to be) supplying them with information.

Which then beggars the question: What kind of information could a seemingly innocuous little professor have that could possibly interest the world's leading intelligence community?*

Others maintain that Fujisawa was working for the Soviets. According to Russell (1992):

> If Fujisawa was a Soviet spy, he was also a distinctly unique breed, one for whom there existed a "transcendent" philosophy in the sense that Japan was seen as a vital link into the "New Age . . ." in this context, both sides (capitalism and communism) could and should be played against the middle, where, in the professor's view, Japan resided like a sleeping tiger.

A curious use of the term, since "sleeping tigers" soon became the nickname for Fujisawa's "harmless" study group/rotating house party where visitors of every ilk would drift in to discuss politics, exchange ideas (and intelligence?), all under the all-seeing eye of the professor.

---

*Remember that 1950s MIT study of Japanese "brainwashing" technique? Part of that information—and perfected technique?—had been gathered by Fujisawa and his colleagues while working as a propaganda specialist for Japan's research department during the war.

Oswald and the Sleeping Tigers

"My hypothesis is that the Japanese themselves cannot be left out of the picture. No previous published research into the [Kennedy] assassination has mined this particular lode. For [Oswald] the influence of the Orient did not culminate in the late 1950s in Tokyo. It continued into the critical years of 1962 and 1963."

In January 1964, Japan joined the investigation into the Kennedy assassination. They wanted to know if there was a Japanese connection to Lee Harvey Oswald, how many people were involved, and what were their ultimate aims? (reported in *U.S. News & World Report*, June 8, 1964).

An Oswald/Japan connection? It's true Oswald had been in Japan. But just how much had Japan gotten into Oswald?

One answer was revealed in 1962 when, after returning from Russia, Oswald was overheard telling a friend he'd first become interested in Communism "thanks to contacts in Japan" (Russell, p. 143).

Lee Harvey Oswald's Japanese odyssey began in 1957 when the young marine radio operator was stationed at Atsugi Naval Base, thirty-five miles southwest of Tokyo.

Atsugi was the perfect place for a man who, barely six years later, would have the entire world tracing and retracing his every step, his every contact, sniffing at and scrutinizing his every sidelong glance for any hint of conspiracy.

In the late 1950s Atsugi base was home to the Joint Tactical Advisory Group (JTAG), the point of origin for U2 spy plane missions.

Atsugi also had the dubious distinction of being the CIA's main base in the Far East, one of two field stations where the CIA conducted secret experimental LSD testing on military personnel (cf. Vankin, 1992).

While stationed at Atsugi, Oswald would often disappear into Tokyo on two-day leaves where he reportedly met with a host of intriguing characters, people from what Russell calls "all walks of the spy spectrum" (Russell, 1992, p. 161).

These were Professor Fujisawa's "Sleeping Tigers." They included Thela, a Soviet intelligence officer Oswald had an affair with, a Soviet colonel named Eroshkin, sundry CIA operatives, intelligence-swapping freelancers, Japanese Communists, and plenty of Asian "poon-tang" able and willing to give a young marine like Oswald not just the time of day—but the time of his life.

One of these, described by some as "Oswald's Eurasian girlfriend," began schooling him in the Russian language.

Oswald was briefly reassigned to the Philippines, but following what appears to be deliberately disruptive discipline problems on his part, his repeated requests for transfer were finally granted and he was transferred back to Atsugi.

It was around this time that Oswald's friends and fellow marines began noticing a change in his behavior. According to one marine acquaintance, Oswald was a "completely changed person, cold and withdrawn, associating more with his Japanese friends and less with other Marines."

This is as good a place as any for us to review some Black Science basics on how to spot the classic signs and symptoms cult indoctrination and brainwashing share in common:

The first step in both is to take the targeted person out of their familiar environment—at least figuratively. Literally removing them from familiar surroundings and familiar cultural support cues (uniform, language, etc.) is best.

He is then exposed to (i.e., bombarded) with a barrage of novel sensory input—new sights and sounds. In the case of POWs, people screaming in a foreign language. In the case of a cult, a lulling language will be constantly "cooed" into the ear of the novice.

He may struggle to learn this new language, be it "cult-speak" or a true foreign language.

In the case of cult indoctrination, the novice hurries to learn the language because the "Master's" teaching and the cult's holy book are both written in that language and, according to the cult, can "only be understood when read in the original language."

This makes the neophyte have to spend much of his time trying to learn a foreign tongue and/or makes him dependent (and subservient) to other members of the cult who do speak a smattering of the "original mother tongue."

Yet even if he succeeds in mastering the language to the point where he can read the Holy Book for himself, the Master still has unique insight into the scriptures, insuring the initiate stays seated at the Master's feet.

In the case of out-and-out brainwashing, e.g., a POW situation, prisoners are often forced to read propaganda out loud—either to avoid "stick" punishment, or in order to receive a reward (e.g., adequate food or medical care)—the "carrot."

Entering into a cult, the neophyte is welcomed with open arms—he leaves his worries at the door. There is free sex, drugs, and rock'n'roll for him to indulge—all without guilt.*

Inside this new circle of "friends," he's repeatedly told how special he is. And since no one else in his life outside the cult seems to realize his special genius, he comes more and more to depend on his new friends for this kind of ego reinforcement.

Eventually, the initiate becomes more and more dependent on the group for his identity. He needs more and more of their constant reassurance that he is special, chosen, one of them.

Soon his former friends and family begin to notice abrupt changes in his behavior. But when he mentions this to his new friends, they explain it away as his old friends being "jealous" of what he's discovered. "The Truth."

Slowly but surely he becomes more and more alienated from his old support network in favor of his new support network.

The same isolation process takes place with any POW foolish enough to accept special treatment from his captors. Accepting special treatment can begin with something as simple as a glass of water or a cigarette, perhaps even promised medical treatment for a wounded comrade. But soon, the "cooperative" POW, even when he thinks he's just "pretending" to go along with the enemy, finds himself cut off from his fellow POWs and inexorably drifting closer and closer to the ideology of his captors.

In both cult indoctrination and prisoner brainwashing, the controller incrementally separates the target from his former support network— friends, family, his old identity.

Fujisawa's philosophy freely mixed international politics with metaphysical musings from Eastern mysticism, primarily Japanese Shinto and Zen Buddhism. This synthesis of both worldly and otherworldly physics and metaphysics succeeded in converting most of his "Sleeping Tigers" into internationalists, attached to no particular camp, addicted only to the teachings of the "Professor."

---

*Charlie Manson bragged that "So long as I control the pussy, I control the men!"

Around this time the professor helped establish the Shinto International Academy for promoting "planetary consciousness" (another oft-used New Age catch-phrase).

In 1959, he published *Zen & Shinto: The Story of Japanese Philosophy*. What was basically a redux of his 1942 book, minus the lavish praise for Nazism and Fascism, but still convinced that it was Japan's destiny to "reconcile" capitalism and Communism, still echoing and extending the Hattori ideal of playing both ends against the middle.

Also in 1959, Fujisawa traveled to the United States where he won converts, teaching Zen and Shinto at New York's New School for Social Research.

He returned to Tokyo in 1960.

After leaving Japan, Oswald defected to Russia, married a Russian woman—Marina—and then defected back to the United States—evidently with few raised eyebrows in the U.S. Intelligence community.

After dabbling in the high-profile albeit short-lived "Fair Play for Cuba Movement"—of which he seems to be the only card-carrying member, Oswald then made a couple of curious trips to Mexico before suddenly abandoning his international life of intrigue to settle down in a quiet Dallas suburb, seemingly content to take a boring job working at a book depository.

Professor Fujisawa died in 1963. JFK too. Oh yeah, and Oswald.*

Fujisawa's timely (?) death didn't mean it was the last time before November 22, 1963, that Lee Harvey Oswald would come under Japanese influence.

A lovely Japanese woman named Yaeko Okui came to the United States in 1959, ostensibly to study *ikebhana* (professional flower arrangement).†

Okui first settled in Dallas, then New York—where you'll recall the professor himself had spent some time teaching?

She finally left New York to resettle in Dallas where Lee Harvey and Marina were introduced to her in December 1962.

---

*Coincidentally, Aldous Huxley, visionary author of *Brave New World* also died in '63. You remember *Brave New World*, the seminal novel about a future where everyone is controlled by daily doses of hypnotic, mind-altering drugs?

†One would think one would have more success studying a Japanese art like ikebhana with acknowledged masters in the country of that art's origin.

Supposedly this was the first time Okui and Lee Harvey had met, though, not surprising in light of subsequent events, some have come to suspect otherwise.

Perhaps it was woman's intuition, or just her in-bred Russian paranoia, but Oswald's wife took an instant dislike to Okui and, in fact, Marina suspected Okui of being some kind of spy (!) and tried to warn her husband (Russell, 1992:162).

Oswald pushed aside his wife's warning and was soon having an affair with Okui, which lead to his and Marina's breaking-up.

After the assassination, Okui was questioned by FBI but released. She returned to Japan shortly thereafter.

The Fujisawa Agenda

"In brainwashing, a fog settles over the patient's mind until he loses touch with reality. Facts and fantasy whirl round and change places, like a phantasmagoria. Shadow takes form and form becomes shadow . . ."
—**Edward Hunter,** *Brainwashing: From Pavlov to Powers,* 1971

We've already established that following the close of World War II the U.S. government began pouring money and personnel into learning more about ancient Asian methods of mind control, including, but not limited to, hypnosis and brainwashing, as well as how modern drugs and technology might be used to enhance both.

Private corporations, most often with the government's blessing, also began investing time and money into uses (and potential abuses) of such methods. This is understandable considering that some of these corporations were already heavily invested in pharmaceuticals and possible mind-influencing technologies, so discovering ancient correspondences and/or compliments to recent R&D breakthroughs could only work to their advantage.

The best example of this (we know about) is the Rand Corporation, who had long been interested in "Influencing Technology" (i.e., mind control).

Rand worked hand in glove with the U.S. government during this time, to the point where it was sometimes hard to tell where the finger ended and the fabrication began.*

As far back as 1949, before the beginning of the Korean conflict—after which "brainwashing" would come into common usage—Rand had issued a cautionary report chronicling Soviet experiments in hypnosis dating back to the early years of Lenin.

The report also suggested what immediate countermeasures the United States should take to close this "mind-control" gap (Russell, 1992:211).

Ironically, the Soviets had been spurred to increase their research into creating remote-viewing "psychic warriors" and better brainwashing methods by rumor that the Chinese were already doing such research (Ostrander & Schroeder, 1970:131).

Ignoring their own warning about Soviet skullduggery, Rand would be one of the first private U.S.-based corporations to undertake trade negotiations with the Soviets barely ten years later.

Rand Development, a subsidiary (sometimes jokingly referred to as a "rogue faction") of the global parent company Rand Corporation, was/is magnanimous in its hiring of former intelligence community personnel, first former OSS agents and then CIA. For example, Rand Development's Washington representative in 1959 was Chris Bird, a psychological warfare specialist and former CIA agent stationed in—you guessed it!—Japan.

Bird, who spoke fluent Russian and Chinese, later became Russian editor/translator for something called Mankind Research Unlimited, an organization Russell describes as "A bizarre Washington think tank that specialized in parapsychology and other behavioral sciences" (Russell, p. 211).

In 1953 it was Richard Helms (who would one day grow up to be the seventh director of the CIA) who proposed the creation of MK-Ultra.

MK-Ultra was the mother (expletive deleted) of all programs when it comes to "Gov-X" mind-control research: "X" for "experimental." This included government-sanctioned research into everything from hypnosis and drugs, to training of remote-viewing "psychic warriors."†

Later, after becoming director of the CIA in 1966, Helms oversaw the

---

*Flash-forward, 2006. Can you say "Halliburton"?

†MK-Ultra high jinks are discussed at length in Lung and Prowant's *Black Science* (2001) and *Mind Manipulation* (2002).

destruction of 152 *MK-Ultra* files dealing with "possible means for controlling human behavior" including radiation, electric shock, psychology, psychiatry, sociology, and what was euphemistically known as "harassment substances," e.g., secretly dousing both civilians and military personnel with LSD. (See "Report to the President" by the [Rockefeller] Commission on CIA Activities within the United States, June, 1975.)

In 1953, before the ink was even dry on the Korean War cease-fire, the CIA had already sent agents to Tokyo to squeeze their Japanese contacts for everything they could dig up on ancient and modern Chinese brainwashing. One of these contacts was Professor Fujisawa.

The following year it was revealed the U.S. government had funded a full study on the feasibility of creating "hypnotized" assassins (Horrock in *New York Times*, February 9, 1978).

And let's not forget that in the 1950s there was that MIT study, which by some accounts relied heavily on research data collected by World War II propaganda expert Fujisawa.

In 1957, the same year Oswald first cuddled under Fujisawa'a black wing, the U.S. Army released a research paper titled "The Hypnotic Manipulation of Attitudinal Effect."

In 1958, the Bureau of Social Science Research (a Rand Corporation subcontractor) issued a report for the U.S. Air Force titled "The Use of Hypnosis in Intelligence and Related Situations."

So was Okui in town simply to "reinforce" some sort of post-hypnotic programming Fujisawa had implanted in Oswald's mind?

Unproven and probably unprovable.

But, at the very least, we owe it to future researchers to diligently catalog our own observations, these curious comings and goings, odd occurrences and highly improbable coincidence, recording them in the same way an ancient astronomer had to be content to record observations made only with the naked eye, content that his data—no matter how paltry—would one day be pored over and appreciated to the point of perhaps inspiring future generations of better-armed seekers to at last uncover the whole truth.

By the way, in that oh-so-well-publicized photograph of Lee Harvey Oswald passing out Fair Play for Cuba leaflets in August, 1963? There's an Asian man standing behind and to his left (noted in Russell, 1992:163).

Even if we do decide to give some credence to the whole Oswald / Japanese influence connection, we're still left pondering as to Professor Fujisawa's motive for creating such an . . . an "Oswald."

To ask why Fujisawa would want to create such an agent beggars the question why would anybody want to create such an agent:

Was Fujisawa under contract to create a "patsy" for (rogue?) elements within the U.S. intelligence community?

Or, all the time Fujisawa was "programming" Oswald (and others?) he was really doing it for his own agenda, unbeknownst to Fujisawa's U.S. handlers?

If that's a "yes," then it leads us back to the question of what precisely was the professor's ultimate agenda?

Let's say he had succeeded in blending ancient Asian mind-control techniques with more modern narco-hypnosis and/or "psychotronics."*

Why succeed in such an undertaking only to then just keep it secret? Possible answers:

- For personal satisfaction. (You know how those "big-head" scientist types are!)
- For pay. Sell an already proven system of mind control to the highest bidder. And what better proof than killing the United States president!
- For payback. In the minds of some hard-line Japanese, the dust will never settle over Hiroshima and Nagasaki.
- Play the super powers off against one another—Hattori again!

Scenario: The Americans uncover evidence Oswald was jacked up by the Russians to kill Jack. Japan then either steps in at the nick of time as mediator. Or else Japan steps back, letting the super powers duke it out, picking up the pieces after Russia and the United States (and hopefully China thrown in for good measure) kill each other off.

Still a little too far-fetched? No proof?

It's not like, down through the years, the uses of hypnosis and other forms of mind control haven't been discussed, discovered to actually work, documented . . . and then those documents just as quickly deleted!

---

*Psychotronics, electronic devices specifically designed to affect the mind.

Yet despite the best efforts of shadowy cadre and self-serving Svengali-like scoundrels to keep such techniques secret from the masses, all the while using those techniques against the masses, still the truth has a habit of leaking out. The best secrets tell themselves. For example, both the defensive and offensive uses of hypnosis have long been established.

The medieval cult of Hashishins (aka the Assassins), the original Muslim terrorists, used hypnosis both defensively and offensively to manipulate the minds of enemy and ally alike.

In the 1960s the U.S. Air Force "Medical Survival Program" taught pilots how to use self-hypnosis defensively to resist brainwashing. In the 1950s *MK-Ultra* programs *Bluebird* and *Artichoke* both experimented with uses of hypnosis.

As previously mentioned, the 1960–1966 Technical Services Division experiments into hypnosis had three goals in mind:

1. Find a way to induce hypnosis rapidly in unsuspecting subjects.
2. Create post-hypnotic amnesia, so the victim would never suspect they'd been hypnotized.
3. Successfully implant post-hypnotic suggestions that the victim would follow efficiently and without question. (Lung, p. 382)

This agenda satisfies the "defensive" use of hypnosis, e.g., a courier could carry information they are unaware they are carrying, and/or "delete" that information if captured.

Such an agent could also be used "offensively" to carry out actions (post-hypnotic suggestions) he was unaware of, actions contrary to his moral compass while in a more "lucid" state.

Such agents would of course be immune to interrogation. And, best-case scenario, would make the perfect assassins, à la *The Manchurian Candidate*.

Drugs ("narco-hypnosis") and/or electronic implants and monitoring devices ("psychotronics") could be used to augment this procedure when necessary.

Once perfected, these same techniques could be (mis)used to discredit politicians refusing to "play ball" as well as snoopy reporters, by inducing schizophrenia-like symptoms.

Then there's the Holy Grail of evil hypnosis: hypnotizing someone to committing murder.

Sure, you've heard the obligatory disclaimer—designed to make hypnotists seem just a little more user-friendly: "You can't make someone do anything while they're under hypnosis they wouldn't normally do."

Heh-heh-heh. Anyone who's studied hypnosis for any period of time (or who reads Dr. Lung books) knows all it takes to make a hypnotized subject commit acts they wouldn't "normally" think of doing is: First, make sure the person is sufficiently "under."

Since there are varying levels of hypnosis—from light trance to deep—and since everyone responds just a little differently to suggestion, it may be necessary to use repeated sessions in order to achieve the depth of trance, i.e., state of deep suggestibility, needed to implant a suggestion-command that might otherwise be resisted during "normal" consciousness.

Second, the real key to convincing your subject to do something he or she normally wouldn't do is to pick the right subject, preferably someone with an obvious glitch in their personality—say, a desire to be famous—and then reinforce in their mind the thought that certain acts will insure their fame—like shooting the president?

Finally, it's all about how you "frame" your suggestion. Rather than tell a person to "Kill Bill!" you have to convince them Bill is an evil dictator, or invading alien, something that will *depersonalize or demonize* Bill to the hypnotized person.

Don't scoff. This is the same thing all countries do when they send their young men (and young women) off to kill people who don't look like them: We depersonalize and then demonize the enemy in order to make it easier for us to blow his "evil" head clean off! It's called "propaganda."

An alternative tact is to convince the hypnotized person to do some "harmless" task, one resulting in the death of the targeted person. For example, have your "patsy" deliver a "birthday present" box (filled with explosives!) to Bill's door. Train the hypnotized person to "play shoot" Bill with one of those toy guns where a "BANG!" flag pops out the barrel . . . and then switch it for an identical (but real!) pistol.

So it can be done. It has been done. We're only left to argue if it was done in Dallas, Texas, on November 22, 1963 . . .

Just how many times does the hydra of hypnotism rear its ugly heads during Oswald's odyssey?

We know Fujisawa mastered ancient Asian techniques of meditation and mind control, adapting them with a New Age spin.

We know the Technical Services Division of *MK-Ultra* field-tested what they called "rapid induction" hypnosis in Mexico City in 1962 and 1966. We know Oswald was in Mexico City in 1962 (Lung, 1997: 382).

We know David Ferrie, a shifty associate of Oswald's during the summer of 1963, bragged of having mastered both psychology and hypnotism. For those of you keeping score, Ferrie reportedly committed suicide.

It gets better. The headliner at Jack Ruby's nightclub the week of the Kennedy assassination was a hypnotist named Bill DeMar, real name William Crowe (Bowart, 1978:213).*

Still not convinced? How about a documented case where hypnosis "brainwashing" was used to create the perfect courier-agent?

Read *The Control of Candy Jones* by Donald Bain (Playboy Press, 1977) where hypnosis was used to turn the model/actress Candy Jones into the perfect courier for the CIA.

FYI: Candy Jones was the wife of talk-show host "Long John" Nebel, who Lee Harvey Oswald telephoned in the summer of 1963 asking to appear on his program (Ibid.).

For those of you who might still think we're just talking out our "tin-foil hat" when it comes to this "Oswald and the Diabolical Japanese Doctor" tale, where one sinister soul can manipulate another hapless one into committing acts that eventually affect the lives of everyone in the world, recall the wisdom of Musashi:

> In my strategy, one man is the same as ten thousand. . . . The spirit of defeating a man is the same for ten million men.

Postscript: "These people play hardball. . . . They want to use drugs for warfare, for espionage, for brainwashing, for control."

(Mary Pinchot Meyer, Washington socialite shot and killed October 12, 1964, by an unknown gunman on a towpath near the Potomac River. Between January 1962 and November 22, 1963, she had had an affair with

---

*And for the real conspiracy buffs in the audience: James Earl Ray visited a Beverly Hills psychologist to be hypnotized to sleep better (Russell, 1992:798).

John F. Kennedy. She spoke these words to longtime friend Timothy Leary shortly before her murder. Timothy Leary is, of course, best known as the "guru" of the New Age LSD movement.)

"The insuperable gap between East and West that exists in some eyes is perhaps nothing more than an optical illusion."
—Pierre Boulle, *The Bridge Over the River Kwai*, 1954

Part 3

. . . AND WEST IS WEST

INTRODUCTION: "THE SWORD IN THE STONE"

Joseph Campbell's *The Hero's Journey* (1990) maps out the quest of the universal hero. According to Campbell, this hero is a universal metaphor found in all times in all cultures, and is designed to remind us of the steps we must take in order to discover our own purpose in life, our own hero within.

Yeah, that's some deep stuff. But thinkers like Joseph Campbell don't swim in the shallow end of the gene pool.

Following Campbell's outline, when we first come across our hero-to-be he's static, beset by doubts in his own destiny, weighed down by a vague sense of dissatisfaction. He'd like his life to change, but doesn't exactly know how to go about initiating that change.

Conveniently (for fictional heroes at least), there's always some outside force that intrudes into the potential hero's life—a couple of star-crossed robots being chased by a galactic serial killer crash on our hero's otherwise peaceful planet. Or an old magician friend suddenly shows up asking him to leave his idyllic village (full of a lot of little people with really hairy feet!), avoid a half-dozen dragons and a hundred other really ugly things trying to kill him, all so he can drop this magic ring into an active volcano. . . . Suddenly plowing that north forty doesn't look all that bad!

Change is hard. Change is scary.

Remember the classic Appalachian American story of an old hound dog laying on the porch moaning up a storm.*

"What's wrong with your dawg?" a neighbor asks the owner.

"Oh, he's amoanin' cause he's a'layin' on top an ol' rusty nail."

"Well why doesn't he just move?" the neighbor wonders aloud.

The owner shrugs, "Reckon it don't hurt him enough to move yet!"

All too often it takes a tragedy to knock us off our familiar ol' rusty nail. And all too often when it comes to that, it's too late to control our life changing in a good way.

Thankfully there's still a few self-starters out there who'll pull themselves up by their bootstraps, gird up their loins, cinch up their gunbelt a notch, straighten their Stetson, and get on with it!

Yeah, change is hard, and there isn't always help on the horizon.

Some look to the East for answers. Some find the answers they're looking for in their own backyard. Some look to the sky; some read Tarot cards and tea leaves; some the latest stock market ticker tape. And some just take a long, hard look in the mirror.

It doesn't matter your particular oracle, it doesn't matter what brand of match you use to light a fire under your ass . . . only that you finally begin to feel the heat!

How can a quest ever be completed if the hero never sets out on that quest in the first place?

In the West, the ultimate quest legend is "The sword in the stone."†

What most people don't realize is that pulling that pig-sticker from the stone wasn't the end of Arthur's journey; it was only the beginning.

So whether you need to pull a magical sword from a stone . . . or just finally move your lazy ass the hell off that rusty nail (!)—a word of warning:

Pull that sword from the stone and your life will never be the same again. Scary, huh?

---

*Appalachian Americans used to be called "Hillbillies," a term no longer considered politically correct. How do you know if you're an Appalachian American? You couldn't spell "Appalachian" to save your life!

†Most know the tale of the sword in the stone from Arthurian legend. Not as many know the much older Norse legend of "The Branstock," a great tree into which High God Odin thrust a magical sword up to its hilt, where it remained until the right hero finally arrived to pull it from the tree. Many experts believe this older legend may have inspired the Arthur story.

Maybe your life will change for the better . . . maybe for the worse. But either way, you can bet there'll be some bumps in the road—a dragon or two along the way to slay, some guy in a black robe (claiming to be your father!) trying to talk you into coming over to the "dark side."

Be a whole lot safer to just leave that sword alone right there in that stone . . . be easier to just keep lying right there on that rusty nail . . . "easier" for your enemies, "better" for those who want to see you fail, those who are even more afraid of you changing than you are of changing yourself.

In fact, you've got a whole lotta people worried sick that if you go gettin' a wild hair to change your life, pretty soon they'll have to change, too, just to keep up.

They worry—and rightly so—that a man who learns the secret of changing just one man (himself) might then know the secret of changing all men. Once he finds out how to change himself—for the better, he might then just decide to change the world—for the better.

First comes the conquest of self—past that point of no return, unknown lands bow before us!

"Self-knowledge and self-improvement are very difficult for most people. It usually needs great courage and long struggle."
—Abraham Maslow

6

Mind Control by the Numbers (Round Three)

"Every human being, is a problem in search of a solution."
—Ashley Montagu

The East is often portrayed (or is that stereotyped?) as being too right brain–dominant, more artistic, less time constrained. Conversely, the West has been accused of being too left brain–dominant, too verbal, too dependent on the power of sheer numbers (firepower?) to solve all the world's problems.

And where we can't find the time (more numbers!) to "crunch" our numbers fast enough, or efficiently enough, we've built a machine to do it for us.

It's true computers give us the promise of one day breaking all of reality—including people—down into manageable (manipulatable!) "ones" and "zeros." But until then, we can still count on numbers (heh-heh-heh) when it comes to factoring ways to first understand, and then undermine, our enemy, organizing our own world in preparation for disorganizing his.

TROUBLE COMES IN THREES

Human beings in general, and Westerners in particular, like it when things come in threes. No one's quite been able to figure out exactly why this is true.

Some developmental scientists speculate that our ever-expanding consciousness is locked inside a slower-to-evolve physical body, itself trapped inside a three-dimensional space.

Others, opting for a more esoteric explanation, claim that man has a threefold nature, a physical, a mental, and a spiritual body.

Still others go all religious on us, admonishing us to remember that it was the Trinity—Father, Son, and Holy Ghost—that literally got this dustball rolling. Of course, thousands of years before the triumph of this cosmic Christian triumvirate over Western thought, the Hindu Brahma, Vishnu, and Shiva—Creator, Preserver, and Destroyer, respectively—ruled first the Indo-Aryan consciousness and then later, with slight modifications, the European psyche.

Even Freud felt the need to organize (or reorganize?) man's troubled psyche into his own version of the Holy Trinity: Id, Ego, and Super Ego.

Later, another psychology pioneer, Fritz Perls of Gestalt, gave us "The Three Types of Shit" patients sling at their therapist. See if any of these sound familiar:

1. *Chickenshit:* small talk, devoid of emotional content and commitment
2. *Bullshit:* lies and fantasies the patient tells about themselves to make themselves look (and feel) more important than they really are
3. *Elephantshit:* grandiose plans, schemes, and impossible dreams the person is definitely going to do . . . someday . . . just as soon as some impossible condition and/or qualifier is met, e.g., when I win the lottery. Hint: "Elephant shit" statements always begin with "When . . ." and "If . . ."

Three's a Crowd

Novice writers are warned not to pack more than three separate bytes of information into the same sentence, since people tend to "zone out," losing concentration after digesting three pieces of information.

Hmmm. Does that mean if we deliberately—albeit surreptitiously—bombard a person with more than three pieces of information at a time, they'll unknowingly process that extra (subliminal) information without conscious thought or censorship?

Kinda like missing the forest for the trees? Missing out on seeing the big picture because we're so caught up in our petty, day-to-day distractions?

Professionally made TV commercials usually stick to this "no more than three pieces of information at a time" rule. In fact most advertisers, not trusting us to remember even three things, take one byte of information and repeat it over and over.

However—and this is where it gets good, good and *sneaky*!—whereas that fast-talking spokesman or narrator on that commercial you're watching is only spitting three bytes of information in your direction *what's going on in the background?*

Next time you're watching your favorite commercial, the one where your favorite basketball star casually stops by in his customized Bentley to shoot some hoops with some thugs from his old neighborhood while one and all sing the praises of their favorite new diet cola or tennis shoe, make sure you take a look at the graffiti-decorated wall behind them.

I know you're not naïve enough to think the producers of that million-dollar commercial just happened to pick any wall on just any basketball court in the inner city?

Long before filming actually begins on such a commercial, professional graphic designers and artists are brought in to *redesign* and *repaint* that backdrop; experts who choose their specific colors and abstract "graffiti" to please (i.e., subconsciously attract) the specific types of consumers the product is being marketed to. And yes, you can bet your bottom dollar—because advertisers invest millions of *their* dollars!—that surveys and research studies have been done out the wazoo to determine what colors and what symbols the targeted audience is most attracted to.

And don't even get me started on how much thought goes into the *music* being played in the background, as well as the T-shirts and other "gear" the "spectators" in the commercial are wearing (each one of them a walking "billboard" for more subliminal advertising).

Now here's the kicker: Remember that "3" is the magic number when trying to put across a message . . . that's three *conscious* bytes of information.

Beyond these three conscious bytes of information, your brain subconsciously "absorbs" the extraneous information without question, promising itself it will "sort out" the extra info later—something our brain seldom gets around to doing.

Thus, all that subliminal "background chatter" goes straight to the "buying" centers of your brain. Gotcha!

And what about the spokesmen in the commercial? A man dressed like Uncle Sam gives you the allowable three bytes of information. That's three bytes of info your brain has just taken in, right? Wrong. Uncle Sam is the fourth subliminal piece of info that you take in, without conscious thought, straight to the abstract section of your brain. Say the spokesman is a mega-sports star or well-known actor. You hear his three bytes of information, but *he* (or *she*) is the real message: drink this soda pop, buy these shoes, wear this brand of underwear and you'll be like me—adored, handsome, happy, successful, and getting laid every night by fans and groupies!

Third Time's a Charm

Ever look at something, somebody, and then have to "do a double-take," taking a second look?

Sometimes the world takes a second look at us—most times, not. That "first impression" rule pretty much holds up under interrogation.

(By the way, you never get a second chance to make a really good first impression—dent, that is!—in your enemy's head, so make that first swing count!)

I'm sure you remember us discussing the three perspectives from which the world views us. These three perspectives influence what we think about ourselves and what others think about us—taking us back yet again to Sun Tzu's "Know yourself, know your enemy."

In life in general, let alone specific, potentially costly undertakings, we must correctly (1) assess these three perspectives and (2) do our best to align these often wildly varying perspectives so that, at the very least, they don't interfere with our schemes—best-case scenario, our reconciling the three ends up assisting us. Failure to do so means the worst kind of failure.

The Way We See Ourselves

> **"Lying to ourselves is more deeply ingrained than lying to others."**
> **—Dostoevsky**

In the East they say, "The eye cannot see itself." In the West we say, "The skunk can't get past his own ass long enough to smell the rest of the world."

In other words, it's hard to make a realistic—ruthless—self-assessment. (And we don't have to. .... Only if we want to survive!) If you don't take the time to ferret out your weaknesses, rest assured your weasel of an enemy will.

The Way Others See Us

Our enemies have a habit of underestimating us. Friends and family overestimate us, often out of kindness—never realizing the danger they are placing us in.

Remember that both your friends and your enemies, like people everywhere, "filter" their perceptions of you through any of a number of personal and social prejudices.

Our friends and family love us . . . that's why, most of the time, we can't trust them to tell us the hard, cold facts of life. Nothing is harder than telling a loved one the cold, hard facts of life. Just remember, the reason they call them the cold and hard facts "of life" is because they're hard and cold enough to be real. It's the "real" that keeps you alive.

Other times it's a good thing to let friends, family, and others sing our praises—free propaganda.

Don't despair. Whether our enemy underestimates us or overestimates us, either can work to our advantage. If our enemy underestimates us, if he thinks us weak—undecided, understaffed, unarmed—he might decide now's a good time to attack us straightaway. If it turns out he has indeed truly underestimated us (because we have successfully camouflaged our true strength), then his campaign is doomed from the onset—faulty intelligence having written his epitaph.

Need we reiterate how it is often to our advantage to deliberately make an enemy underestimate us, in order to entice him into overextending himself and lead him into a trap? Think Custer. Think Trojan horse. Think Sun Tzu, "When strong, appear weak . . ."

If, on the other hand, our forces are truly weaker than his, we need to disguise that fact as well, letting him see only what we want him to see. The face we wear to win the world is not the face we'll wear to rule the world.

The Way We Really Are

The way we really are falls somewhere between how we see ourselves and how others see us.

Nowadays there's so much of that Dr. PhilOprahDonahueWilliams, "Feel good about yourself no matter how fucked up you are," "accept yourself as you are," crap floating in the toilet bowl of life that we're actually beginning to believe it's all right to say anything, do anything, to anybody you want—sans consequence.

The truth is most people don't know how messed up they really are. And they're never gonna figure it out if self-promoting, would-be experts on human behavior keep assuring every numb-nuts, child-molesting, drug addict that "I'm okay, you're okay."

Well, I *am* okay. But I don't know about you. And there's a good chance, unless you're one of my close friends or family, that I don't really *care* about you. So there's not much chance I'm gonna open my hermetically sealed bomb shelter door for you when those Islami-fascists finally set off the Big One—or even the Little One.

You're not responsible to save everyone in the world. In fact, not everybody is *worth* saving. Somebody's gotta feed the worms.

On the other hand, if you can convince me (through three inches of steel) that helping you and yours survive will help me and mine survive . . . the key's under the "Welcome" mat.

Suppose you take a long hard look at yourself, factoring in how other people see you as well, and then decide you want to change yourself?

You might decide to change yourself simply because you don't like the way you see yourself, or because you don't like the way others perceive you (and subsequently act toward you).

Simply by employing the Five Warning F.L.A.G.S. we can make people "see" us differently. (See page 226.)

Anytime we choose, we can re-make ourselves, making significant and permanent changes that will affect the rest of our lives . . . or else just change ourselves temporarily, slapping a smiling false face on long enough to talk that wishy-washy customer into buying that previously owned car.

It's not that difficult to make people see what you want them to see when they are looking at you. The key is figuring out what it is they *want* to see in you.

THE FIVE WARNING F.L.A.G.S.

FLAG	GESTURE/ACTION	HOW THEY PERCEIVE YOU
Fear	Threatening words & actions; "leaked" information about past acts of violence *you've committed*, real or fabricated for affect.	They perceive you as "a dangerous man". Reputation often spills less blood. Lesser mortals kowtow at your feet, joining your cause.
Lust	*You confide* your licentiousness to another in order to draw their "confession" out. You fan the flames of their fetishes for YOUR blackmail files.	They trust you as a kindred soul (i.e. just as big a *pervert* as they are!) and they confide their secrets to you, secrets you "bank" for later.
Anger	*We show* impatience and exasperation/disappointment (heavy sigh!) with a colleague/co-worker, staying vague as to the specific cause of your anger.	Confused, they compensate by trying all the harder to please you, to assuage their "guilt".
Greed	*You act* very "secretive" about a "special" deal you've got in the mix, whettng their appetite by your furtiveness. You "accidentally" leave your prospectus open for them to "see".	You have the inside track. You know something they don't. If they invest their money in you they will get rich! Cult variation: You have God on speed-dial.
Sympathy	*You make a show* of your generosity & "caring", shaming them into following suit to match your "compassion".	Shame-faced, they try to match your generosity. You help them donate their time & money . . . to YOUR cause!

Figure 24.

Consider: A first time car buyer walks onto a car lot scared and paranoid—scared he or she is going to make mistakes, paranoid some stereotypical used car salesman is going to high-pressure them into buying a clunker.

What this person is *looking* for is a friendly, helpful, knowledgeable face. *Be* that person and you'll make the sale.

What's that beautiful, albeit middle-aged woman sitting at the bar looking for? That's what you have to figure out so you can *become* the lover she's looking for. (See *"The Art of Seduction"* section in my book *Mind Control*, 2006.)

As we'll see in a minute, what people are "looking for" in life can be summed up into three main mind-sets, or motivations: Pleasure, power, or meaning.

Once you peg the person you've targeted as having their head in one of these three motivations, it's a simple matter for you to "shapeshift" your strategy and your tactics (i.e., yourself) to fit their mode of motivation, changing how they see you by changing who you are, or at least who you appear to be.

Sometimes people don't know what they're looking for, so they probably won't know what they are looking *at* (you) until you tell them what to *see*, and thus what to *think* about you.

You can accomplish this easily enough by studying the sections that follow on "Shapeshifting".

Any time *you* decide, *you* can change how people "see" *you*.

Granted, it's easier to change how people perceive you from positive to negative (just fart enough) than it is to change people's perceptions from negative to positive (saving a litter of puppies from a burning building . . . that is, unless you're the one who started the fire in the first place!).

But it can be done. Has been done. And it will be done—by you—if your need, or desire, is strong enough.

Change begins and ends with motivation. Yours and your target's. No one changes without the proper motivation. Sometimes a person has a need, a drive within them, usually fueled by a specific goal, that motivates them to change.

Other times, depending on your own motivation, *you* provide the motivation for your target to change.

Pain is a great motivator. So are the Five Warning F.L.A.G.S.

Some folks are inner-directed (those annoying "self-starters" again). Others need a reason to even get up in the morning—like Joseph Campbell's universal hero who wants to change his life but probably never will until something comes along to jar him out of his malaise.

Sometimes you have to be your own drill sergeant. At other times, all of us need a kick in the pants to get started, a little help from outside ourselves. But only initially.

Sooner or later you'll have to learn to stand on your own two feet, that is if you want to be able to knock your enemy off his!

Three on a Match

In psychology there are three main theories for what motivates the psyche of man.

Freudian Theory

Named of course for Grand-daddy Sigmund himself (1856–1939), this theory says that man is motivated by a "will to pleasure." We seek those things that are the most pleasurable to us, avoiding those things that bring us displeasure and pain.

Curiosity, confusion, and conflict (not to mention the occasional indictment!) occur within self and between self and the world—when your idea of pleasure and society's idea of pleasure are two different things. This is especially true when your idea of pleasure somehow involves automatic weapons, confections from Colombia, and small farm animals. . . .

Adlerian Theory

Named for Austrian psychiatrist Alfred Adler (1870–1937), Adlerian Theory maintains that what drives man is not pleasure per se, but the "will to power," the need within man to be the master of his domain, cock of the walk, to exert force and to see the fruits of that force manifest in his life.

Yeah, this does sound a lot like the same thing Friedrich Wilhelm ("God is dead get over it!") Nietzsche (1844–1910) had said some years before, except that, when Nietzsche said it, it sounded a whole lot more "take over the world"-ish!

Frankl Theory

Also known as Logotherapy, founded by Viktor Frankl, this theory disputes both Freud and Adler by giving a more optimistic assessment of man. Frankl (1973) states that man's motivation is the "will to meaning," the desire to understand and bond with the universe around him:

> Man's search for meaning is a primary force in his life . . .
> This meaning is unique and specific in that it must and can
> be fulfilled by him alone. (p. 154)

228

There is nothing in the world, I venture to say, that would so
effectively help one to survive, even the worst conditions, as
the knowledge that there is a meaning in one's life. (P. 164)

Interestingly, whereas Adler agreed with Nietzsche that man's primary
marching order is the "will to power," in his 1959 *Man's Search for Meaning:
An Introduction to Logotherapy,* Frankl also quotes Nietzsche:

He who has a *why* to live can bear almost any *how*.

According to Frankl, both Freud's "will to pleasure" and Adler's "will to
power" exist, but are of a lower order than the "will to meaning." That's
why the "will to meaning" when frustrated vicariously compensates by
becoming the "will to power" (seeking money, position, etc.) and/or else
drowns its misery (his failure to achieve meaning) in the "will to pleasure"
(e.g., sex).

So what we have here are the three major motivating forces that drive
us: pleasure, power, and meaning.

These stack up pretty well against Abraham Maslow's famous "needs"
pyramid that begins with our satisfying our physical needs and (hopefully)
culminates with our achieving our "self-actualization" needs. (See figure 12.)

Rather than embracing one of these three to the exclusion of the
remaining two, in effect limiting ourselves—never that!—Black Science takes
the view that we are all, to varying degrees, susceptible to all three of these
motivations.

Let's assume for a minute Frankl is right that both "will to pleasure" and
"will to power" are subservient to the more noble "will to meaning." Even
if that's true, *especially* if that is so, we can use this knowledge to our advantage to disadvantage our foe.

If we deliberately frustrate our enemy's "search for meaning" in his life,
following Frankl's formula, then we can easily tempt him further—confusing,
corrupting, and ultimately crushing our little puppet-to-be by offering him
pleasure and power—*strings* firmly attached!

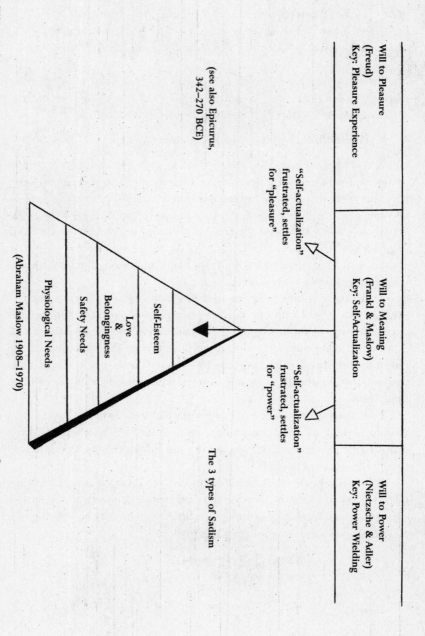

Will to Pleasure
(Freud)
Key: Pleasure Experience

Will to Meaning
(Frankl & Maslow)
Key: Self-Actualization

Will to Power
(Nietzsche & Adler)
Key: Power Wielding

(see also Epicurus, 342–270 BCE)

"Self-actualization" frustrated, settles for "pleasure"

"Self-actualization" frustrated, settles for "power"

The 3 types of Sadism

Self-Esteem

Love & Belongingness

Safety Needs

Physiological Needs

(Abraham Maslow 1908–1970)

Figure 25.

230

How to Be a Sadistic S.O.B.

(What Would the Marquis de Sade Do?)

> *"Between the promise and the performed, between the expected and the actual, falls the shadow of deviance."*
> —**Lyman & Scott,** *A Sociology of the Absurd,* **1989**

The Marquis de Sade (1740–1814) spent thirty years in prison, including the final thirteen years of his life.

And let's face it, history is not going to take you seriously as a writer or as a philosopher after your name becomes synonymous with tying people up and torturing them! (I know, I know, what two—or more—consenting adults do in the privacy of their own sound-proof basement dungeon/recreation room with their whips and chains and imported lubricants is their business.)

But, as you know by now, Black Science students go out of their way to find those misunderstood, much maligned, and occasionally maniacal masters of mind manipulation who have been cursorily passed over by the more conventional—respectable?—sciences of the mind.

That means, rather than confront our enemy with strategies, tactics, and techniques that anyone with a library card can check out, we're coming at him with tactics torn from forbidden folios, and with plots and ploys prized from the priestly pariah and the maligned Master alike.

To our enemy's shock and awe, we've armed ourselves with mental machinations and mind-blowing maneuverings thought long lost, revelations on human behavior that become our battle plan for total mind penetration.

Ours is not the politically corrected, sanitized, and homogenized pabulum taught the herd. Our texts are those banned long ago—and tomorrow, scribblings and scrolls long thought burned and buried. And those they're gathering even now for tomorrow's bonfire of the vanities. Ours are the writings of men who saw too much and talked too little, quiet men finally made silent by small minds crowded with giant fears and petty hates.

You can bet our enemy's never seen anything like what we're bringing to his table!

Sometimes the obscure is the perfect weapon with which to obscure your foe. And you can bet no one—with the possible exceptions of Judas, Hitler, and Osama bin Laden—has been more maligned than the Marquis de Sade.

Like Machiavelli and Freud, de Sade is condemned because he dared rip aside the black curtain of conformity, propriety, and poseur etiquette to expose the orgy of hypocrisy hidden beneath.*

Admittedly, most of de Sade's writings are *still* considered scandalous 200 years later—even by today's standards. However, many of his more insightful and scathing commentaries—attacks!—on the nature of man in general, and the decline and fall of his own French society in particular, have been lost, ignored, or deliberately suppressed. Suggested reading: *Dialogue Between a Priest and a Dying Man.*

But even in his most licentious works, *Justine, 120 Days Of Sodom,* de Sade, an embittered veteran of the Seven Years' War (aka, the French and Indian War), still finds time to throw barbs at what he saw as the cause of France's decline: the corrupt aristocracy and the perverted priesthood. (A corrupt government and pedophile priests? Times certainly have changed!)

It had been argued that within de Sade's philosophy all three of our previous theories for the motivation of man—Freudian, Adlerian, and Frankl's—come together.

Even with the traditional concept of physical "sadism" we find elements of all three of these motivations: The Sadist finds pleasure in being sadistic (Freud's will to pleasure), he or she can derive a feeling of power (Adler's will to power), and he might even find a meaning for existence (Frankl's will to meaning).

Indeed, to some extent or the other we are all "sadistic," at least according to the definition given by Stanford M. Lyman and Marvin B. Scott in their mind-blowing *A Sociology of the Absurd* (General Hall, 1989).

Before you dismiss this idea off-hand as preposterous, consider that most people don't know, or else don't care, that there are actually two types—or at least "degrees"—of sadism, the study of which can give the Black Science student a glimpse into potential plots and ploys utilizing this knowledge.

Don't panic, only one of these categories of sadism actually involves whips and chains and desperately trying to remember your "safety word."

---

*At one point in his life, the Marquis was in fact sentenced to Death in Absentia in Marseilles, condemned as much for his offensive writings and for his infamous reputation, as for actually taking "liberties" with a young lass.

Sadism Proper

> **"Sadism covers an enormous range of human activity. From the creation of works of art to the blowing up of bridges, from making little girls happy by giving them sweets, to making them cry by slapping them."**
> —**Geoffrey Gorer,** *The Life and the Ideas of the Marquis de Sade*, **1962:156**

Lyman and Scott define sadism proper as "the pleasure felt from the observed modifications on the external world produced by the will of the observer" (*A Sociology of the Absurd,* 1989:193).

On the surface this sounds a lot like Adler's (and Nietzsche's) "will to power," that need we have to feel we have some kind of control, or at least influence, over our lives.

This is what's known as our "power base," our "comfort zone," and contributes to our overall "support network."

This need to feel "in control" is inherent in all people and is linked to Maslow's "survival and safety needs." (See page 230.) Realizing this, gives us a vital clue for (1) understanding ourselves better and (2) undermining our enemies better.

Since we all like to feel we have some control over our lives, the Black Science adept approaches his target with promises of either increasing that person's power base, or else with threats of taking away what power the target presently possesses. Thus we comfort him, making him our friend and ally (or, at the very least, our bitch!), or else we discomfort him—knocking him out of his comfort zone, and taking away his power base—weakening him for the coup de grace.

Possessing a strong power base and support network makes us feel safe—insulated and happy. At this point Sadism proper dovetails into Freud's "will to pleasure." Lyman and Scott (1989) add:

> The pleasures of sadism do not spring from the unhappiness of others, but from the knowledge that one is responsible. According to British anthropologist Geoffrey Corer, what is central to sadism is the pleasure one receives by knowing that one's words or actions affect others strongly. . . . Sadism is the pleasure a bank robber has when he looks at the horrified face of a teller as he says, "Stick 'em up!" Sadism is giving a

233

great performance on the dramatic stage to the applause of an audience. In short sadism involves any perceivable modification of the external world through one's own efforts. If sadism connotes destruction and cruelty, it is because one is more likely to perceive a modification of the world by blowing up a bridge than by building it. (pp. 193–94)

According to the Marquis, the "sadistic" urge (need) is natural, man's most basic instinct.

Toddlers fearlessly explore their new world, teenagers deliberately fly in the face of convention—testing themselves as they test the limits of tolerance and gravity.

By the time we're adults, for most, this natural instinct has been bred (or beaten) out of us. Still we love Freddie Kruger movies, jumping out of airplanes, climbing mountains, and we cheat on our wives—all of which are designed to raise our stress levels; thrill-seeking behaviors, all actions that can result in potential disaster if we lose, a feeling of power—of being alive!—if we succeed.

And, oh yeah, men love war.

It's no secret deliberately raising our stress levels—in a controlled situation, or at least in a time and place of our own choosing—brings us pleasure.

In this way, Sade's philosophy resembles Freud's "will to pleasure."

There are two distinctions (or degrees) to sadism proper: the more passive Schadenfreude, and the decidedly more aggressive Algolagnia.

Schadenfreude

Schadenfreude is the sense of joy and/or excitement we get when we watch others suffering pain and unhappiness. But while we enjoy this other person's pain and discomfort, we are not the ones directly (actively) responsible for their misfortune and suffering (Lyman & Scott, 1989:193).

I can hear you protesting now, "Okay, I would never—could never!—take joy in watching another person's pain or discomfort!"

Ever laugh at someone who's slipped on a banana peel? What about laughing while watching one of those "hidden video" programs where they put people in the figurative hot seat?

"Oh but that's different," you argue, "nobody ever *really* gets hurt on one of those shows!"

Okay, aren't you going to be out dancing in the streets when they finally get around to sticking Osama bin Laden's head on a pole?

"Oh, but that's different, too. He's just getting what he's got coming to him . . . that's justice."

So that's a "Yes"? You *will* take pleasure in seeing ol' camel-face suffer?

Schadenfreude pops up in two ways: benign and belligerent.

Some manifest Schadenfreude through more benign, *positive* actions—or, more often, their *lack* of action.

When something bad happens to someone, it's karma, justice, fate, kismet. It was meant to happen, universal recompense for past deeds. You reap what you sow, that sort of thinking.

This type of Schadenfreude mind-set takes joy when bad things happen to bad people because it's just the universe righting itself, bringing itself back into balance, a form of cosmic restitution.

The more belligerent side of Schadenfreude looks through more *negative*-tinted lenses and sees retribution and revenge. Like their more benign brethren, they also see the suffering of others as karma, the bad guys getting their just deserts. But unlike their brethren, these guys feel it's their job to lend a hand to justice, giving the universe a hooded helping hand, helping restore order. God's little helpers, what would he do without them?

Vigilantes and religious fundamentalists often share this mind-set:

"I am not Karma, but I can be her sword!" (Joshuah Tree)

These range from the completely *passive*, non-interfering, to *passive-aggressive*, affecting conditions through their inaction or "slow-walking."

Most often this mind-set is content to just sit back and watch events unfold—even when a word from them could prevent a disaster.

This type of person is guilty of lies of omission, and is the best friend Gossip, Propaganda, and Conspiracy ever had. He is the consummate snitch.

Algolagnia

> *"Although psychoanalytic literature often identifies algolagnia with sadism, it might be more properly thought of as a specialized type of sadism."*
> —Lyman & Scott, 1989:193

235

Finally, we get to the whips and chains. Algolagnia is defined as the pleasure (including sexual pleasure) you get from deliberately and actively inflicting pain on others.

Unlike Schradenfreude, Algolagnia is active, not passive, and is in the job of purposely causing pain.

Thus, within sadism we can see the distinct personality types: sadism proper, and its two properties: Schadenfreude (passive to passive-aggressive), and Algolagnia (aggressive).

Once we figure out which one of these personality types best fits our target, our enemy, we can then use the inherent weaknesses within each to dominate him (see page 237).

"How could I possibly hurt a fellow human being without also hurting myself as well? Wear Thick gloves, stupid!"
—Duke Falthor Metalstorm

Extremes of either pleasure or pain are dangerous to society, especially the more totalitarian that society. Too much pleasure and we find respite and escape. Too much pain and we learn to endure long past what our oppressors are capable of meting out, thus the bastards lose their only power—fear!—over us.

WHEN YOUR NUMBER'S UP

"Fear not, gentlemen, we shall always have passions and prejudices, since it is our fate to be subject to prejudices and to passions."
—Voltaire, *Philosophical Dictionary*, 1764

Over 240 years ago, in his *Philosophical Dictionary*, Voltaire (1694–1778) made the observation that man is subject to four types of prejudice:

1. *Prejudice of the senses:* Our Senses deceive us at every turn. (That's why *we* took up the Red Spears' Jing Gong senses training earlier.)
2. *Physical prejudices:* We are betrayed by a body beset by fears and lusts; constantly and consistently confused by the coquettish, coaxed to go off half-cocked.

SADISM

The pleasure felt from the observed modifications on the external world produced by the will of the observer
—Lyman & Scott, 1999:193

Type of Sadism	Mind-Set	Approach/Attack
1. Schadenfreude a. Passive: Observing others' suffering and unhappiness brings us pleasure, though we are not the direct cause. Examples: A comedian slips on a banana peel; an evildoer gets caught and executed. b. We allow someone to suffer through our inaction and/or lies of omission.	Let's things unfold "naturally," doesn't interfere. Cold and calculating. Voyeur. Live and let live (or let die!). Vicarious thrills. God's little helper. Helps restore balance to the universe through non-interference. Plausible deniability in all things. Secretly a coward. Afraid of commitment.	Agree with him that "people get what they deserve." Advise him to wait, don't get involved. Give him all the excuses he needs, plausible deniabilty. Become his enabler. Lessen his guilt for not acting ("bank" his inaction/ complicity for later blackmail use). Turn him into an informer since he loves gossip. Use him to spread dis-information.
2. Algolagnia Aggressive. We derive pleasure from being directly responsible for another's suffering.	Lies of commission. Troublemaker, likes to stir the pot. Closet bully. Miserly. Power seeker. Scared of pain and death so he hides behind others, seeking from their pain and fear how to avoid his own.	Assist his assaults and indiscretions (keep a file for later use—don't forget your own alibi!). Encourage him to overextend himself. Help him "cross the line" of no return . . . without your help, that is.

Figure 26.

237

Functioning at this bestial level, we think only of flight-or-fight, or the occasional bite of forbidden fruit.

3. *Historical prejudices:* Corrupted (and corrupt*ing*) cultural myths and self-serving societal superstitions that bind us to traditions outdated, and manipulate us through patriotisms long since played out.

4. *Religious prejudices:* Revelations and rants carved in cold, unforgiving stone, reinforced further by black-robed robber barons whose hearts were cut from that same queer quarry. We dare not question these proclamations from on high, or at least from those who sit on high thrones; their fane P's & Q's thrust so deeply into us at so tender and so trusting an age.

These are the four kinds of "filters" through which we sift all the information we encounter in the world, almost from the moment we are born and their indelible number is carved into our forehead, to the moment our number's up.

The Four Types of Mind Games

"Do not reveal your thoughts to everyone, lest you drive away your good luck."
—Apocrypha, Ecclesiasticus 8:19

So far we've learned that people differ when it comes to which of the five senses "dominate" them at any given time and which of the five, in turn, predominate them most of the time. So, too, when it comes to playing "mind games," it turns out there are four distinct types of "mind games" people play when trying to get one up and one over on their fellows.

Which one of these mind games, or combination of mind games, one uses depends in large part to what "body orientation" controls them, or at least influences them the most. Thus, "mind games" people play fall into four types: head games, face games, heart games, and groin games.

Head Games

We need intelligence to function, both the innate kind and the gathered variety. Head games specialize in the gathering of information—overtly and covertly. We then either "bank" the information we've gathered for use at a later date, or we spend it freely to help us accomplish our immediate goal.

Head games have two objectives: First, gather the information we need to accomplish our objective, storing it inside our own head till needed.

The second objective of head games is to literally get inside our enemy's head, to find out (1) how he thinks, (2) what he thinks about, and (3) how we can influence (and ultimately *control*) how he thinks and what he's thinking about.

Once we uncover these three tidbits of knowledge, and any other piece of information that might aid us in achieving our agenda, we can then incorporate that intelligence into the three other types of mind games, mixing and matching tactics and techniques until we arrive at our intended goal.

It's simple: I want to know information you don't want me to know, and vice versa. It's a game of spy-versus-spy that ultimately comes down to "seekers and leakers: two roles we all play at one time or another depending on which is more advantageous to us at the time.

"Seekers" seek information, looking for "the edge"—that insider stock market tip, the combination to the safe, what time your husband's *not* gonna be home.

"Leakers" are those giving out the information, often selling the information. Sometimes leakers leak information accidentally (or with secret encouragement from us), other times they give out information subconsciously.

Someone (a leaker) has information. Someone (a seeker) wants that information. This is the universal dance that keeps the universe going, or at least our little ball of dust spinning.

This formula applies whether you are talking about stealing nuclear secrets, or stealing a kiss from that hot piece of eye candy just down the bar.

You have information, I want it. I'm willing to pay for it. What do you want for it? Money. A Godfather-favor to be cashed in sometime in the future?

Maybe you've got some kind of psychological need that needs paying— an apology for a real or imagined wrong done you. Maybe you just like to see people grovel before you tell them what they want to hear?

I wash your back, you wash mine. Just don't drop the soap because not everyone operates on the honor system!

Any interaction between two human beings, between groups, between corporations and even countries, follows this basic give 'n' take setup.

Bottom line: we are seeking compliance (we either want information or we want someone to do a certain action for us) and we are prepared to get it by any means necessary.

In a perfect world, we achieve this goal with the least amount of effort and expenditure possible. The goal is the most amount of gain for least amount of pain—to self or others.

To accomplish this, we have to make the other guy see things *our* way; whether it's something small like accepting our filling station's inflated price for gasoline, or something on a grander scale, like accepting that our Big Duke country will nuke the hell out of your little puke of a country if we don't get what we want!

Our head game is designed to alter another's mind, to control their thinking, if only long enough to make our sale—*and* cash the check.

This is the core of any kind of salesmanship, whether selling a used car or selling yourself.

Our techniques include looking for patterns in behavior—in order to upset those patterns, to throw the enemy off his precious schedule, to make him rush headlong into a blunder.

We spy out the enemy to uncover what he prefers to keep hidden. Once we gain the upper hand, we must then take steps to guard against his ever (re)gaining the upper hand over us.

Thus we must remain wary of revealing too many of our moves and mind games to our enemy. We show him only those moves necessary to insure his demise! And, if we're really good, he never even sees those coming!

Face Games

Face games are all about establishing and maintaining a secure identity. By "secure identity" we mean an identity—a mask, if you will—that serves your needs, whether those needs be immediate or long-term.

This identity needs to be accepted (authenticated, validated) by those who will most come in contact with it.

In the case of a short-term identity (e.g., one adopted say to run a "short-con") your fard may need only fool the lazy security guard at the door.

For a more long-term commitment, for example, the more genuine "face" you show your family, your coworkers, and the world in general on a daily basis, in effect who you are, this "face" cannot appear inconsistent.

Society tolerates a variety of roles, roles we are all asked to choose from, in order to help perpetuate society. This is not necessarily a negative thing. These accepted roles each carry with them certain rights and responsibilities, as well as rewards for performing them dutifully for the promotion of society as a whole.

Thus, during your lifetime, you may take on the role of dutiful son, slightly rebellious teenager, dutiful husband, father-figure to your children, solid citizen, soldier, and deacon in your church—all roles necessary for the maintenance of society. Correctly perform these roles and society will reward you with platitudes and plaques.

Of course you are also free to choose the roles (identities) of bully, criminal, ex-con, anarchist; negative roles, yet all of which also serve society, if only by validating those others who have assumed the roles of police, judge, lawyers, jury members, prison guards, and—increasingly—executioners.

Other are necessary to help us maintain our chosen identities by agreeing to "recognize our face."

These others validate our identity, especially when it is in their best interest to do so. Often validating our identity lends more credence to the identity (false face?) they are trying to establish and/or maintain.

Validation from others can be either beneficial or toxic, it can lead to us becoming better, more productive human beings, or else thrust us into the thralldom of dependency on others for our feelings of self-worth.

Such toxic "dependency" (aka "co-signing") is at the rotten core of all cult relationships: I acknowledge (and never question!) the "Master" and, in turn, the "Master" gives me my validation, reinforcing my identity as a loyal, useful cult member. This identity is further validated by my fellow cult members. They validate me, I validate them, the "Master" validates us all. (They used to call this kind of thing "a circle jerk"!)

And while we're on the subject of manipulating roles and "identities" for personal gain, recall from my book *Mind Control* (2006), "The Art of Seduction" section how "scoring" is all about validating (i.e., pretending to buy into) what kind of identity (role- playing mask) she's wearing tonight. She got her cowgirl "face" on, she's lookin' for a cowpoke, and you better be packin' a six-shooter!

All societies are built and maintained around mutual agreement as to the value placed on identity and "status"—in this case, the two being synonymous.

We don't greet the president of the United States with "Yo, Georgie!" It's just not done. We all stand when the judge enters the courtroom. Violate the "chain-of-command" in the military and you're asking for trouble. And even if we don't subscribe to a certain religion, we still give our "respect" (i.e., validation) to members of their clergy.

Face games are the glue (actually more like spit and bullshit!) that holds society together.

The breakdown of society begins when "respect" (i.e., co-signing/validation) for leaders and institutions begins to fray around the edges. If everyone doesn't agree to play by the rules . . .

During the sixties, various hippie, yippie, and yahoo groups went out of their way to challenge and even openly "disrespect" established political titles and entitlements and "stuck in the mud" social institutions they saw as standing in the way of progress.

Remember, identities—face games—only work when the other guy plays along.

So far as Black Science strategy is concerned, you have to establish your "identity" early on in any situation in such a dramatic way as to allow no one to "challenge" your claim to that identity.

Often, as with cult leaders, this is done by validating the faux face of those around you.

Suppose you want to start your own cult?

You can start from scratch—making up your own "Bible," rituals, etc., or you can join an already recognized and established religion: infiltrate this already established group, isolate those vaguely dissatisfied and/or bored members, and slowly turn the group to your way of thinking.

This can only be accomplished if you can (1) manipulate the already existing identities of the members and/or (2) manipulate your own identity so that you will appear as their figurative (and sometimes literal) savior; perhaps the one come to explain the "true meaning" of scriptures they've been reading for years?

You accomplish this by showing them novel ways to interpret existing scripture (while sneaking a few of your own self-serving psalms and proverbs in from time to time).

With each such "tweaking" of their already established belief, you subtly (re)mold *their* identities until they become dependent on *your* identity (as "Master") and start coming back to you again and again, ego-junkies now hooked on you to validate their new identities.

In the East, they have the concept of "saving face." "Face" being loosely translated as "respect" and your "standing in the community."

You never deliberately make another person "lose face," i.e., disrespect them by not respecting (validating) their position in life. That is, unless it's your enemy and you want him to "lose face," thus undermining his influence and authority.

Your enemy will then either have to retire in shame from the playing field (leaving you the victor by default), or else he will be forced to rash and imprudent extremes in an effort to regain his face.

The master Black Science strategist would have, of course, anticipated this latter course of action on the part of his disgraced enemy. In such an instance, following Sun Tzu's teaching to always leave an enemy a way out, we proffer the perfect way our enemy can regain face—actually a way to lead him to further ruin, or else force him to put himself at our disposal.

In the West, in popular vernacular this is called "fronting someone off" or "pulling their hole card," exposing them for the poseur they are.

Of course, if you're the only one who knows someone is "putting up a false front," faking a more colorful past, laying claim to accomplishments and awards they don't really have, you can "bank" that information for use later in blackmailing—uh, I mean "negotiating"—with them.

Heart Games

As the name implies, heart games deal with affairs of the heart, relationships—establishing them and maintaining same, either to accomplish a purely exploitive agenda or else for a more altruistic motive.*

Heart games are also used to end relationships, to either break it off or cause the other person to break it off. Sometimes people do this subconsciously, doing stupid things to (deliberately) sabotage their relationships because of a "fear of intimacy and/or commitment." (See, it does pay to stay awake in Psych 101!)

---

*Recall Black Science Rule #256: There's no such thing as true altruism—everybody gets paid, whether in earthly ducats or heavenly deductions.

Heart games cover all human interaction, not just romantic interaction; family interaction, business interaction, and societal interaction in general.

What distinguishes heart games from head games and from groin games is *emotional investment*.

Heart games take a lot of emotional investment to achieve, to maintain, and, all too often, to terminate.

Manipulation-wise, the more emotion we inject into our enemy's mind, into our plots and ploys against him, the better our chances of flustering and frustrating him, and ultimately causing his own plans to go completely catiwhompus.

The things that affect us the most are the things that tug (and sometimes tear!) at our heartstrings. The simple explanation for this is that while we process much of our interactions with others in the higher (logical/reasoning) part of our gray matter, our emotions are still centered in the antecedent "reptile brain." That's why emotional things literally affect us on four "deeper"—literally, deeper inside the brain—levels.

Groin Games

Groin games are not just confined to, and concerned with, our nether regions. Groin games are "influence" games, a jockeying for the three P's: Position, Power, and Prestige.

If you think you're hearing echoes of Nietzsche and Alfred Adler, you are. When playing groin games we're exercising our "will to power": seeking *power* (within our self and within the world) and *prestige* (recognition and validation) in the eyes of our fellow man—and fear in the eye of our enemy! This is all aimed at our better securing a *position* higher up on the unforgiving food chain of life. The groin games we play help us achieve this coveted, comfortable position, safe and secure from those lowlifes still fighting for scraps at the bottom of the evolutionary heap.

Groin games are ultimately all about survival. And that means we use what we have to get away, get our way, get ahead, and ultimately get back at anyone who ever got in our way by trying to play *their* mind games on *us*!

Groin games allow us to mix 'n' match techniques from the previous games.

All the mind games people play come down to two variables: (1) defining the situation (i.e., setting the stage) and (2) establishing identities (ours and the other guy's).

Think of this as setting up a play. We either set up a situation (encounter), with an idea—already in our mind—of how that situation should play out, or else we walk onto a stage where a play is already in progress, a play set up and scripted by someone else.

It is at this point we decide either to "play along" or to disrupt this play already in progress, replacing it with a script of our own.

When we set up our own play (e.g., a job interview, a con game, a romantic seduction) we choose the setting in the same way a director sets the stage for the production of his play, the way a wily general always picks his battlefield beforehand.

We then assign "roles" or identities to the people who will be acting on that stage. You will play the lead (the con man, the seducer, the used car salesman), while you assign others the role of "victim" or "supporting cast."

Sometimes this works out and everybody agrees they like your choreography . . . other times, *your* version of how the play should go is challenged.

Plays—like life in general—only work when everybody "agrees" to go along, i.e., "follow the script."

Ultimately, whether you're talking about a specific kind of play, or life in general, it's all about whose definition of "reality" wins out, who's got the upper hand? It's all about whose definition of "setting" and "identities" wins out—whose playbook or script are we gonna go by?

The good news is you can manipulate these two variables ad infinitum. The bad news is, so can your enemy. So you have to watch your back, 'cause no matter how good an "actor" you are, there's always some eager-beaver understudy waiting in the wings for you to stumble. (Of course, the really ambitious—or just impatient—understudy *helps* you stumble.)

On the great casting couch of life . . . it's better to be the director.

Recap: In all the games people play someone has something someone else wants. If the other guy's got something *we* want, we try to convince, cajole, coerce and, as a last resort, cudgel him into giving it to us.

If on the other hand *you* have it (whatever "it" happens to be—car, career, or just *cootchie*?), you advertise. You get the word out. You get potential customers all hot 'n' bothered.

You wave "it" in front of your bull-headed enemy like Superman's cape.

You pique his curiosity, stir his interest, you hook him with a good offer . . . then you pull the ol' "bait 'n' switch" and he ends up paying big time.

245

You charge whatever the market will allow . . . and then try to squeeze just a little bit more out of him. What he imagined was a gold ring, ends up being a ring made of brass, one that fits perfectly in his nose!

Whatever it is you have to sell, you have to sell yourself first; all the while keeping an eye out for your competition, and for whatever line of bull your enemy might be trying to sell *you*!

Once you figure out what kind of mind game your enemy is playing, you can counter him with one—or more—mind games of your own; mixing and matching, baiting and switching, till he doesn't know if he's coming or going and doesn't see your ploy coming until you're going out the door with everything he holds dear! (See p. 247.)

Another "Four" for Honorable Mention: Machiavelli's "Four Elements" in the "Putting Your Mach' Hand Down" section that follows.

Paracelsus' Five Causes of *Dis*-ease

"People heve neglected to study the secret forces and invisible radiations."
—**Paracelsus**

Philippus Theophrastus Aureolus Bombastus ab Homhenheim, thankfully better known as "Paracelsus" (1493–1541), lived with one foot in magic the other in the science. Struggling to free himself from medieval superstition, all the while Paracelsus was forced to kowtow and cater to lesser minds, lest he suffer the same fiery fate of other early scientific masterminds—martyrs like Giordano Bruno.

But fear was never enough to stop Paracelsus. Continually seeking after knowledge, a practical physician always on the lookout for new cures, Paracelsus was also an insatiable natural scientist, attacking problems the way a warrior hurls himself into the heart of a battle:

"Every experiment is like a weapon which must be used in its particular way—a spear to thrust, a club to strike. Experimenting requires a man who knows when to thrust and when to strike, each according to need and fashion." (Paracelsus)

Paracelsus lived—and did his best to keep living!—back when authorities (i.e., the Church) burned first and asked questions later, when a mere

THE FOUR TYPES OF MIND GAMES

	Goal	Approach
Head Games (Information)	Information a. Gathering b. Use and distribution	He has information you desire, use the "Killer B's." Intellectual level attacks: Make him dependent upon you for his information. Feed him disinformation. Control his information flow.
Face Games (Identity/standing)	Seek to establish identity to increase standing/position in community and with others	Ego level attacks: Bolster his inflated ego, "co-sign" for his identity to others, then he will become dependent on your continued lying for him. Save him from an embarrassment you arranged in the first place.
Heart Games (Relationships)	Establish and maintain meaningful relationships: a. Mate relationships b. Family c. Society/business Secondary: maintain control over relationships	Emotional level attack: Find out what they need, what they are looking for in a relationship (*Mind Control*, p. 141, "The Art of Seduction").
Groin Games (Power and Pleasure)	Logistics: Secure an endless supply of power and pleasure. Climb Maslow's pyramid.	Body level attack: Security needs. Offer him power and guilt-free pleasure. Help him justify and rationalize his infidelities and ruthlessness. Give him permission to be selfish. Put him to sleep by awakening his needs.

Figure 27.

allegation of suspicious activity or sacrilegious thought was enough to get you invited to a Catholic cookout.

Prudently, Paracelsus practiced medicine openly, while studying metaphysics and protoscience like Alchemy in secret:

"For good reasons did the ancient magicians express their prophecies in images rather than in writing. For who dare tell the naked truth to a king? I'd rather not—my reward might be hanging. No magus, astrologer, or chiromancer should tell his sovereign the naked truth. He should use images, allegories, figures, wondrous speech, or other hidden or roundabout ways." (Paracelsus)

Despite his precautions, like all thinking men of his time, Paracelsus was closely watched by the authorities, and on more than one occasion openly accused of *nigromancy* (black magic). Only his incessant wanderlust kept him one foot ahead of the faggot's flame.

In 1518 he began a trip around the known world, reportedly journeying as far as the Orient:

"No man becomes a master while he stays at home, nor finds a teacher behind the stove. Diseases wander here and there the whole length of the world. He who would understand them must wander too." (Paracelsus)

Paracelsus was not your soft, sterotypical scientist born with a silver test tube in his mouth. His early life had been arduous, toughening him, fortuitously preparing him for many hardships—and many hasty exits!—in later life:

"Through his life, a man cannot cast off that which he has received in his youth. My share was hashness, as against the subtle, prudish, superfine. Those who were brought up in soft cloths and by womenfolk have little in common with us who grew up among pine trees." (Paracelsus)

He spent much of his time treating—and carousing with—common people, 'cause that's where all the really interesting diseases were! As a result, many of his friends were poor villagers, "villeins," from which we get our modern word "villains."*

---

*For more on how this and other "curse words" came into being, see the sections in *Mind Control* on "Word Slavery" and "Word Weapons."

In fact Paracelsus by all accounts spent as much time in the tavern as he did in the laboratory. In other words, he couldn't confine himself to just book learning, so strong was his natural desire to learn and empower himself:

"To be taught is nothing; everything is in man, waiting to be awakened." (Paracelsus)

You learn more from correctly reading one man's face than you do from reading a hundred other men's books.

Paracelsus' hearty and unconventional life is also reflected in his untimely death at a robust forty-eight.

Paracelsus' official date of death was recorded as September 24, 1541.

One version of his death has him succumbing to an overdose of laudanum, which he always kept hidden and handy in the pommel of his sword. Some sources refer to the laudanum as a nasty, personal habit Paracelsus picked up during one of his many travels; hence his death could have been accidental.

Others claim he always carried just enough laudanum with him at all times to commit suicide, in case he was ever in imminent danger of capture—and most assuredly torture—by the Church.

The fact that he carried it in the pommel of his sword, the fact that he felt the need to go around armed in the first place, tells us a lot about his unpredictable lifestyle—or at least something about his paranoia.

Still others claim Paracelsus died during a drunken tavern brawl. Adding conspiracy theory spice to this already juicy version of his death are allegations that the brawl was sparked by, and Paracelsus ultimately killed by, assassins sent by one of his many enemies.

Master alchemist and physician, deserving an honored place in both the medicine and chemistry halls of fame, Paracelsus helped set the standard for scientific enquiry and experimentation, beating back the darkness of ignorance with the light of research and reason.

One of Paracelsus' most insightful discoveries was his distinguishing "the five causes of physical diseases." According to Paracelsus, there are five *"entia"* (principles) with the power to influence and distress mankind: the stars, man's body itself, his humors, his mind, and Acts of God.

But on a deeper level of understanding, Paracelsus was also speaking of the *"dis-ease"* of the mind. This was dangerous ground to tread back in those days.

Like most medieval masters, Paracelsus was often forced to hide many of his insights and discoveries into human nature, for fear of inciting and igniting the torches of those who believed anything above man's "gross" physical nature to be their exclusive purview, if not their private playground.

Whereas Church authorities tolerated a physician's (even an *alchemist's*) blood-letting cures for the physical body, something they themselves could benefit from, above and beyond that you were stepping onto sacred and superstitious ground where it was believed matters, maladies, and madness of the mind could only be controlled and/or cured by the purification of a man's morals: those morals, in turn, to be decided on and dictated by The Church and her secular servants.

By the Stars (ens astri)

> **"Paracelsus denied that the stars influence man directly, and he laughed at the idea that disease, too, have a horoscope."**
> —**Henry M. Pachter,** *Magic into Science: The Story of Paracelsus,* **1951**

While the stars are the first "influence" Paracelsus listed, it is interesting to note that he himself didn't believe in astrology.

So how *do* the stars cause or otherwise influence our *dis-*ease? By our believing in them. Recall from *Black Science* (2001) and *Mind Manipulation* (2002), in the sections on "The Twelve Beasts" of Junishi-do-jitsu (Asian astrology), it doesn't matter if *you* believe in astrology (or any other kind of voodoo for that matter), it only matters that *your enemy* believes in it; that he believes in it either because he genuinely feels it will help him, or else because he *fears* you've got the "Mojo" and will not hesitate to use it against him.

Knowing what our enemy believes gives us a heads-up on which way he's going to jump when we hook our Black Science electrodes to his testicles.

Hitler, and many in his inner circle, believed in astrology. That's why British military intelligence had to hire their own astrologer—so their astrologer could read Hitler's chart every morning and tell British Intelligence what Hitler's astrologer was telling him, things like "Your sun-sign is rising in Scorpio so today would be a good day to invade Poland, mein Fuhrer!"

The problem, dear Brutus, is not in our stars, but in ourselves. But if our enemy believes the problem is in *his* stars, it's just one more superstition we can use against him.

By the Body (ens veneni)

We are slave to the body. It gets sick, it gets old. It breaks down. Sometimes it foolishly helps hasten its own breakdown by ingesting heaping helpings of drugs and alcohol and by having sex with hookers who ingest heaping helpings of drugs and alcohol . . . a skank who just happens to be shacking up with a recently paroled biker boyfriend waiting for her back at the trailer park who has just discovered his personal heap of drugs and alcohol is missing and your name is on his bitch's speed-dial!

Some are born with defect and deformity to the body, others receive these courtesy of a drunk driver or Uncle Sam. Injuries that cripple the body also cripple the mind.

Some of us have phobias and allergies, both of which work against us— both of which can be turned against us by an alert foe (see *Mind Control,* 2006, pp. 20–21).

By His Humors (ens naturale)

The concept of the "Four Humors" originated with Sicilian philosopher Empedocles (490–430) theorizing that the four elements (Earth, Air, Fire, and Water), thought to compose everything in the universe, are found in the human body as four body fluids: Phlegm, Blood, Choler (aka Yellow Bile), and Black Bile.

Empedocles' Four Humors theory continued to dominate medical thought in Europe for the next thousand years, up into the Renaissance, Paracelsus' time.*

Good health comes from maintaining a reasonable balance of these four humors. Conversely, physical illness, often called "Ill Humor," comes from us having too much of one humor, too little of another.

---

*"The Four Humors" are discussed at length in *Mind Control* (2006) in the chapter of the same name.

Mental illness (mental *dis*-ease) is brought about—and can be cured—the same way, by rebalancing the Four Humors.

By discovering which of these humors is out of balance (i.e., which one is missing), the physician can prescribe medicine, herbs, etc., which are infused with (i.e., contain) some of that missing humor, thus restoring the patient's balance.

Of course, for the reprobate-minded in our class, one could easily do the same thing, only in reverse—throwing your enemy's humors out of balance by introducing various chemical potions, charms, and even specially crafted words and phrases (i.e., spells) each corresponding to a specific element/humor.

CONCLUSION

*"The great leaders of the human race are those who have
awakened man from his half-slumber. The great enemies of
humanity are those who put it to sleep, and it does not matter
whether their sleeping potion is the worship of God
or that of the Golden Calf."*
—Erich Fromm, *Beyond the Chains of Illusion*, 1962

If the choice is between God and the Golden Calf, choose the one that gives
the most milk.

GLOSSARY

ASP: "Additional Sensory Perception." The full use of our five senses that give the impression to others we possess a "sixth sense", i.e., "ESP."

Assassins: Medieval Middle Eastern Muslim secret society cult noted for its terror, treachery, and mind-manipulation techniques.

Awfulizing: Imagining the worst that can happen, making mountains out of molehills. Chronic worry.

Banking: Holding back valuable and/or damaging information (indiscretions, faux pas, etc.) you've discovered about a person for use in blackmailing and/or disgracing them at a later date.

Big Brother: Oppressive government, always watching. Coined by George Orwell in his 1948 novel: *1984.* (See "Orwellian")

Bio-Resources: People whose talents you can utilize to accomplish your goals.

Black Curtain, the: Generic, the veil of secrecy and skullduggery sinister cadre hide behind. Synonym for "smoke screen." Specific, the head of a Japanese Yakuza crime family.

Black Science, the: Generic, any strategy, tactic, or technique used to interfere with and/or undermine a person's ability to reason and respond for themselves.

Bloodtie: Dangerous and damaging information we hold over another. (See the "Killer B's")

Cheng and Chi: Chinese, "direct" and "indirect" strategies. Also spelled Zhing and Qi.

Cock-Blockin': Generic, deliberate, or inadvertent interference in the plans of another. Specific, interfering with the seduction plans of another person.

Cogniceuticals: Drugs designed to enhance or entrance the mind.

Cult-Speak: Special passwords and coded phrases cults and cliques use to identify one another while marginalizing "outsiders."

Droppling Lugs: Using innuendo and rumor to plant doubt and seed suspicion, especially intended to undermine another's credibility.

Dyshemism: Words used as weapons. (See "Word Slavery")

ESP-ionage: Research and/or application of "Extra Sensory Perception" to gather intelligence, e.g., when spying.

Fard: Literally, "to paint the face with cosmetics." Generic, to wear a false face, to assume a false identity, aka "False Flaggin'."

Five Warning F.L.A.G.S., the: The five Gojo-goyoku weaknesses: Fear, Lust, Anger, Greed, and Sympathy.

Gojo-Gyoku: See "The Five Warning F.L.A.G.S."

Gov-Speak: The Capitol Hill run-around aka "spin."

Half-Assin': Hesitation and second-guessing. Exactly what we want our enemy doing!

Huta: "Head up their ass." Someone who doesn't want to know the truth and/or is too lazy or corrupt to seek out the truth; such slugs are sometimes referred to as "belonging to the HUTA tribe."

Illuminati, the: Generic, the ultimate secret society bugaboo and boogeyman. Whispered about for centuries, The Illuminati reportedly controls the world economy and pulls the "Strings" of world politics from behind the "Black Curtain." Specific, secret society in Bavaria circa 1776. (See also "Nine Unknown Men")

Isaacs: Anyone who follows orders/authority without question. From the "Isaac Effect."

"Killer B's," the: Techniques for infiltrating as enemy's mind: Blind; Bribery and Blackmail; Bloodties; Brainwashing; Bully; and Bury.

Kyonin-no-jutsu: Japanese Ninja art of using an enemy's superstitions against him.

Mekura: Japanese, the "inner eye," i.e, insight and intuition.

Mindwar: Preemptive measures (propaganda, etc.) used to attack an enemy's mind, intended to sap his will to fight *before* physical war becomes necessary. Sun Tzu's ideal.

"MK": Spook-speak for "mind control." Coincidentally, these same initials are used to identify the MERCK pharmaceutical company rumored responsible for helping government agencies develop cogniceuticals. (See "Spook-speak")

Nine Unknown Men, the: (1) Used as a euphemism for the "Illuminati," (2) multi-cultural myth-legend of nine enlightened "Masters" who walk the earth at any given time. When one dies, another takes his (or her) place. In some versions, these nine rule the Illuminati.

Ninja (Japanese, "to steal in"): Assassin-spies originating in medieval Japan, known for their stealth and skullduggery. Generic (small "n"), anyone who employs stealth and secrecy to accomplish their ends.

Orwellian: Totalitarian, intrusive invasion of privacy by "Big Brother."

Plausible Deniability: Spook-speak for being somewhere else when the fecal matter collides with the oscillating rotor. *Propaganda*: Rumor's big brother.

Propheteering: The Cult Game. Generic, hiding behind religion for deceitful and devious purposes.

Psychotronics: Any electronic device used to enhance or entrance the mind. (In 1970s Czechoslovakia "psychotronics" was used as a synonym for "parapsychology." [Ostrander & Schroeder, 1970])

Shadow-talk: See "Tells."

Shadow-walk: See "Tells."

Siddhas (Sanskrit, "powers"): Enhanced powers of mind and body claimed by Hindu yoga mystics and fakirs.

Spook-speak: Euphemism and code words used by intelligence agencies.

Suggestology: The science/art of suggestion. Includes and/or touches on hypnotism, the power of persuasion, propaganda, etc.

Synarchy: Rule by secret societies, pulling the strings from behind the scenes.

Tantric (Sanskrit, "forbidden"): Taboo mystical practices (drugs, sex, nigromancy, etc.) used by Hindu mystics as a shortcut to enlightenment and siddhas.

Tells: Twitchin', itchin', and bitchin' body language and speech faux pas that inadvertently reveal what a person is *really* thinking and/or may reveal a person's unconscious desires and fears. Also known as "Shadow-talk" and "Shadow-walk."

Thought Reform: Brainwashing by any other name.

"3-D": What propaganda does to an enemy: Demeans, Dehumanizes, and Demonizes—all in order to demoralize.

Word Slavery: The deliberate use of words and language to control and/or otherwise influence another human being. Includes the use of subliminals, culturally taboo words, slur-words (insults) and purr-words (lulling and soothing words).

"X": Spook-speak/Gov-speak for "experiment."

Zetsutjim (Japanese, "offspring of a talkative tongue"): An accomplished talker and manipulator. A Mastermind.

AUTHORITIES

Bayer, Ronald. *Homosexuality and American Psychiatry: The Politics of Diagnosis.* New Jersey: Princeton University Press, 1987.

Bhagavad-Gita (The Song of God) (Misc. translations)

Bocking, Brian. *A Popular Dictionary of Shinto.* NTC Publishing Group, 1997.

Bodansky, Yossef. *Terror! The Inside Story of the Terrorist Conspiracy in America.* SPI Books, 1999.

Bowart, W. H. *Operation Mind Control.* New York: Dell, 1978.

Bryner, Michelle. *"Some Pregnancies Are Not So Accidental."* Psychology Today (September/October 2005):32.

Davis, Erik. *Technosis: Myth, Magic and Mysticism in the Age of Information.* New York: Harmony Books, 1998.

Dhammapada (Sayings of the Buddha). (misc. trans.)

Estabrooks, George. *Hypnosis.* 1948.

Fujisawa, Chikao. *On the Divine Mission of Nippon: A Prophecy on the Dawn of a New Age.* 1942.

————. *Zen and Shinto: The Story of Japanese Philosophy.* Westport, CT: Greenwood Press, 1959.

Goldberg, Philip. *The Babinski Reflex.* California: Jeremy P. Tarcher, 1990.

Holzer, Robert D. *ESP and You.* New York: Hawthorne Books, 1966.

Horrock, Nicholas M. *"CIA Documents Tell of 1954 Report to Created Involuntary Assassins."* New York Times, February 9, 1978.

Hucker, Charles. *China's Imperial Past.* Stanford, CA: Stanford University Press, 1975.

Huff, Darrell. *How to Lie with Statistics.* New York: W.W. Norton and Company, 1954.

Hunter, Edward. *Brainwashing: From Pavlov to Powers.* New Jersey: The Bookmailer, 1971.

Icke, David. *Alice in Wonderland and the World Trade Center Disaster.* Bridge and Love Publishing, 2002.

Kaplan, David, and Alec Dubro. *Yakuza*. Addison-Wesley Publishing Company, 1986.

Krishnamurti, J. *The First and Last Freedom*. 1954.

Kruglinski, Susan. *"What You See Is What You Don't Get."* Discover (February 2006):13.

Laqueur, Walter. *Terrorism*. Boston, MA: Little, Brown and Company, 1977.

Lawson, Willow. *"Battle on the Car Lot."* Psychology Today (September/October 2005):28.

Ledeen, Michael A. *Machiavelli on Modern Leadership*. Truman Talley Books/St. Martin's Press, 1999.

Lepp, Ignace. *The Art of Being an Intellectual*. Trans. Bernard Murchiand. New York: Macmillan Company, 1968.

Lung, Haha. *The Ancient Art of Strangulation*. Boulder, CO: Paladin Press, 1995.

————. *Ninja Craft*. Ohio: Alpha Publications, 1997a.

————. *Assassin! Secrets of the Cult of the Assassins*. Boulder, CO: Paladin Press, 1997b.

————. *Knights of Darkness: Secrets of the World's Deadliest Night-fighters*. Boulder, CO, 1998.

————. *Cao Dai Kung-Fu*. Port Townsend, WA: Loompanics Unlimited, 2002.

————. *Theatre of Hell: Dr. Lung's Complete Guide to Torture*. Port Townsend, WA: Loompanics Unlimited, 2003.

————. *Assassin!* New York: Citadel Press, 2004a.

————. *Knights of Darkness*. New York: Citadel Press, 2004b.

————. *Lost Fighting Arts of Vietnam*. New York: Citadel Press, 2006a.

————. *Mind Control*. New York: Citadel Press, 2006b.

Lung, Haha, and Christopher B. Prowant. *Black Science: Ancient and Modern Techniques of Ninja Mind Manipulation*. Boulder, CO: Paladin Press, 2001.

————. *Mind Manipulation: Ancient and Modern Ninja Techniques*. New York: Citadel Press, 2002a.

————. *Shadowhand: History and Secrets of Ninja Taisavaki*. Boulder, CO: Paladin Press, 2002b.

————. *Ninja Shadowhand: The Art of Invisibility*. New York: Citadel Press, 2004c.

Lung, Haha, and Eric Tucker. *The Nine Halls of Death: Ninja Secrets of Mind Mastery*. New York: Citadel Press, 2007.

Lyman, Stanford M., and Marvin B. Scott. *A Sociology of the Absurd* (2nd Edition) New York: General Hall, 1989.

Machiavelli, Niccolò. *The Prince*. 1513. (misc. trans)

————*The Art of War*. 1520. (misc. trans.)

————Discourses. 1531. (misc. trans.)

Musashi, Miyamoto. *Go Rin No Sho (A Book of Five Rings)*.(misc. trans)

Omar, Ralf Dean. "Ninja Death Touch: The Fact and the Fiction." *Black Belt* (September 1989).

————. *Death on Your Doorstep*. Ohio: Alpha Publications, 1993.

————*Prison Killing Techniques: Blade, Bludgeon and Bomb*. Port Townsend, WA: Loompanics Unlimited, 2001.

————. *Blood on the Sidewalk*. Ohio: Alpha Publications, publication pending.

————. *Steel Nation*. Ohio: Alpha Publications, publication pending.

————. *Direct Action: Take Back the Streets!* Ohio: Alpha Publications, publication pending.

Only, Joshuah. *Wormwood: The Terrible Truth About Islam*. Ohio: The Only Publications, 2006.

Ostrander, Sheila, and Lynn Schroeder. *Psychic Discoveries Behind the Iron Curtain*. New Jersey: Prentice-Hall, 1970.

Packard, Vance. *The People Shapers*. New York: Bantam Books, 1977.

Pauwels, Louis, and Dr. Jacques Bergier. *The Morning of the Magicians*. 1971.

Pickett, Lynn, and Clive Prince. *The Stargate Conspiracy*. New York: Berkley Books, 1999.

Ramanujan, A. K. *Speaking of Siva*. New York: Penguin Books, 1973.

Ringer, Robert J. *Winning Through Intimidation*. New York: Crest/Fawcett, 1975. Reprinted 1993.

————. *Looking Out for # 1*. New York: Crest/Fawcett, 1977.

Russell, Dick. *The Man Who Knew Too Much*. New York: Carroll and Graf Publishers/Richard Gallen, 1992.

Sargent, William. *Battle for the Mind: A Physiology of Conversion and Brainwashing*. 1959.

Seagrave, Sterling. *The Soong Dynasty*. New York: Harper and Row Publishing, 1985.

Shah, Ikbal Ali, editor. *The Book of Oriental Literature*. New York: Garden City Publishing Company, Inc. 1938.

Shaver, Kelly G., and Roger M. Tarpy. *Psychology*. New York: Macmillan Publishing Company, 1993.

Skinner, Dirk. *Street Ninja: Ancient Secrets for Surviving Today's Mean Streets*. New York: Barricade Books, 1995.

Skloot, Rebecca. *"Can Memory Manipulation Change the Way You Eat?" Discover* (January 2006):30.

Soren, David, et. al. *Carthage*. New York: A Touchstone Book/Simon & Schuster, 1990.

Starr, Douglas. *"Animal Passions."* Psychology Today (March/April 2006):94–98, 100–101.

Stine, Jean Marie. *Double Your Brain Power*. 1997.

Sun Tzu. *Ping-Fa (Art of War)*. (misc. trans.)

Tuchman, B. W. *Stillwell and the American Experience in China 1911–45*, 1970.

Vankin, Jonathan. *Conspiracies, Cover-ups and Crimes: Political, Manipulation and Mind Control in America*. New York: Paragon House, 1992.

Voltaire. *Philosophical Dictionary*, 1764. (misc. trans)

Walker S.O.J., and J. Leslie. *Nicollo Machiavelli: The Discourses*. Original translation 1929.

Weise, Elizabeth. *"Men, Women: Maybe we ARE Different . . ."* USA Today (August 22, 2006):9D.

Wetter's, Ethan. *"Why Do People Behave Nicely?"* Discover (December 2005):37–41.

Wilson, Colin. *Rogue Messiahs*. 2000.

Yourcenar, Marguerite. *Mishima: A Vision of the Void*. Trans. Alberto Manguel, 1986.